PRACTICAL HANDBOOK

of

Spreadsheet Curves
and
Geometric Constructions

Deane Arganbright
Mathematics/Computer Science
Whitworth College
Spokane, Washington

CRC Press
Boca Raton   Ann Arbor   London   Tokyo

LIMITED WARRANTY

CRC Press warrants the physical diskette(s) enclosed herein to be free of defects in materials and workmanship for a period of thirty days from the date of purchase. If within the warranty period CRC Press receives written notification of defects in materials or workmanship, and such notification is determined by CRC Press to be correct, CRC Press will replace the defective diskette(s).

The entire and exclusive liability and remedy for breach of this Limited Warranty shall be limited to replacement of defective diskette(s) and shall not include or extend to any claim for or right to cover any other damages, including but not limited to, loss of profit, data, or use of the software, or special, incidental, or consequential damages or other similar claims, even if CRC Press has been specifically advised of the possibility of such damages. In no event will CRC Press's liability for any damages to you or any other person ever exceed the lower suggested list price or actual price paid for the software, regardless of any form of the claim.

CRC Press SPECIFICALLY DISCLAIMS ALL OTHER WARRANTIES, EXPRESS OR IMPLIED, INCLUDING BUT NOT LIMITED TO, ANY IMPLIED WARRANTY OF MERCHANTABILITY OR FITNESS FOR A PARTICULAR PURPOSE. Specifically, CRC Press makes no representation or warranty that the software is fit for any particular purpose and any implied warranty of merchantability is limited to the thirty-day duration of the Limited Warranty covering the physical diskette(s) only (and not the software) and is otherwise expressly and specifically disclaimed.

Since some states do not allow the exclusion of incidental or consequential damages, or the limitation on how long an implied warranty lasts, some of the above may not apply to you.

IBM PC® is a registered trademark of International Business Machines Corporation; Lotus 1-2-3® is a registered trademark of Lotus Development Corporation; Macintosh® is a registered trademark of Apple Computer, Inc.; Microsoft Excel®, Word®, and Windows® are registered trademarks of Microsoft Corporation; PostScript® is a registered trademark of Adobe Systems, Inc.; Quattro Pro® is a registered trademark of Borland Corporation; WordPerfect® is a registered trademark of WordPerfect Corporation.

Library of Congress Cataloging-in-Publication Data

Arganbright, Deane.
   Practical handbook of spreadsheet curves and geometric
constructions / Deane Arganbright.
     p. cm.
   Includes bibliographical references and index.
   ISBN 0-8493-8938-0
   1. Curves—Graphic methods—Data processing. 2. Electronic
spreadsheets. I. Title.
QA483.A74 1993
516'.15—dc20   93-7960
                                                                                                                                          CIP

This book represents information obtained from authentic and highly regarded sources. Reprinted material is quoted with permission, and sources are indicated. A wide variety of references are listed. Every reasonable effort has been made to give reliable data and information, but the author and the publisher cannot assume responsibility for the validity of all materials or for the consequences of their use.

Neither this book nor any part may be reproduced or transmitted in any form or by any means, electronic or mechanical, including photocopying, microfilming, and recording, or by any information storage and retrieval system, without permission in writing from the publisher.

Direct all inquiries to CRC Press, Inc., 2000 Corporate Blvd., N.W., Boca Raton, Florida 33431.

© 1993 by CRC Press, Inc.

International Standard Book Number 0-8493-8938-0

Library of Congress Card Number 93-7960

Printed in the United States of America 1 2 3 4 5 6 7 8 9 0
Printed on acid-free paper

PREFACE

The electronic spreadsheet is one of the most productive and widely used items of applications software for personal computers. In addition to being extremely effective for business and economic modeling, the spreadsheet is also a natural tool for an extensive range of creative mathematical modeling applications. Furthermore, modern spreadsheets possess outstanding and powerful graphical capabilities that are easy to use effectively. This book is designed to bring the power and excitement of the spreadsheet's intrinsic graphic strengths to such fields as mathematics, science, and engineering.

The book discusses the fundamental concepts of spreadsheet graphics for mathematical applications. It also presents a comprehensive assortment of models that are designed to demonstrate how the mathematical and graphical features of spreadsheets can be exploited to generate a diverse variety of classical mathematical curves and to implement creative techniques for geometric constructions. Numerous illustrations produced by these models are also displayed. The book furnishes computer users with new opportunities for employing their own spreadsheet programs to design an unlimited variety of useful and eye-catching curves and illustrations.

FEATURES

Versatile Applications of an Ubiquitous Tool. The models, examples, and concepts that are presented in this book can be implemented on all of the major commercial spreadsheets, as well as on most of the other spreadsheets, without the need for using add-in programs. Thus, inventive graphic ideas and powerful techniques are introduced through a medium that is readily available to everyone for use on their own computers.

Displays of Curves. This book contains many diverse illustrations of curves that are created by spreadsheet models. These curves are even more striking when a spreadsheet displays them in color on a monitor or projection devise. Methods for creating these graphs and then embedding them in word processing documents are discussed in an appendix.

Creation of Interactive Models. Each of the models in this book is designed to utilize the spreadsheet's interactive format through "What if...?" investigations that actively involve users in analyzing various aspects of the curves produced. With a little experimentation, a user can quickly see the effects of changing the parameters of a function that defines a curve, of choosing different fixed points for a geometric construction, or of replacing one underlying curve by another.

Illustration of Mathematical Concepts. Unlike computer programs that simply create curves for a user to manipulate, these spreadsheet examples are designed to involve a user actively in their construction. The models both use and illustrate the underlying mathematical ideas. Thus, creating the models reinforces the concepts involved. Moreover, the spreadsheet approach provides a basis for understanding a large number of geometric construction techniques. This enables the spreadsheet to become a profitable tool both for individual applications and for teaching uses.

New Spreadsheet Graphic Techniques. Many completely new spreadsheet graphing ideas are introduced throughout the book. The techniques discussed significantly expand the range of graphical approaches that traditionally have been carried out on spreadsheets. For example, the XY graph type is utilized to create elaborate mathematical curves and to illustrate functions of a complex variable. In addition, innovative techniques are presented for drawing entire families of lines, circles, and other curves. These procedures are used in creating specific curves, striking designs, and classical geometric constructions.

Animation. Previously unexplored graphical techniques for generating animation effects on standard spreadsheets are discussed. These methods exploit features that are already available on spreadsheets. Chapter 4 highlights many ways to create and use a wide range of animation effects to illustrate features of curves and geometric constructions.

Innovative Graphical Uses of Solver Commands. The equation solver and goal-seek commands are among the newer mathematical features that are available on spreadsheets. Such commands create new mathematical opportunities and applications for use with more complex mathematical models and problems. This book presents previously unexplored uses of these commands for such graphical applications as sketching higher order algebraic curves and illustrating the inverses of complex variable mappings.

Creating Quality Illustrations. Each of the explanatory illustrations displayed throughout this book has been created by these same spreadsheet models. By augmenting the book's examples with a few more rows and columns, spreadsheet users can create diagrams that contain such features as auxiliary lines, identifying text, and the coordinates of points. The resulting figures are updated automatically by the spreadsheet when any one of a model's parameters is altered. In addition to enhancing a spreadsheet's use for personal investigations and professional demonstrations, such graphs can be exported easily to word processing files for inclusion in written documents.

Activities and New Avenues for Creativity. One of the goals of this book is to open new vistas of mathematical and graphical concepts for exploring. In addition to presenting a wide range of examples, this book also suggests new approaches to other ideas to pursue through an extensive range of activities listed at the end of each section. It is further anticipated that a wide range of entirely new ideas to pursue will arise as a reader investigates the topics and models throughout the book.

Disk of Examples. An accompanying disk provides working examples of 25 models selected from the book. These are created in each of the spreadsheets *Quattro Pro* (2.0), *Microsoft Excel* (4.0), and *Lotus 1-2-3* (2.3). They are usable on these and higher numbered versions of the spreadsheets, as well as on *Lotus 1-2-3* (2.2).

AUDIENCE AND PREREQUISITES

The ideas in this book will find applications by professionals in such fields as mathematics, computer science, engineering, and the sciences, as well as in any area that involves the design and use of curves. The examples, concepts, and overall approach also are easily accessible to students in these areas. In addition, those who are interested in finding new avenues and ideas for creating drawings and artistic curves should find useful ideas throughout the book.

While a significant part of this book assumes some background in calculus, many of the examples do not even require this amount of mathematics. Thus, this book will also be attractive to individuals in the vast audience of spreadsheet users who are interested in learning new and creative graphical techniques and applications. No particular previous computer or spreadsheet experience is assumed. An initial chapter that provides a spreadsheet tutorial enables even those individuals with limited computing or spreadsheet backgrounds to use these models effectively.

USES FOR THIS BOOK

Professional Reference. This book is designed to be useful as a professional reference book for mathematicians, computer scientists, engineers, and scientists. Not only does it provide illustrations of extensive and diverse sample curves and geometrical constructions, but it also shows how to create them. The book enables a spreadsheet to be used as a practical and powerful device for creating illustrations of curves for use in professional publications and for interactive demonstrations. Moreover, it is anticipated that the book will generate among readers in a variety of disciplines new ideas for spreadsheet graphical applications, and that some of these ideas will ultimately result in professional publications.

Teaching. The book can make an excellent text for either a special topics course in curve constructions or for students who are interested in an independent study. The book is designed so that the ideas within it can be pursued by students largely on their own. There is wide flexibility in the order in which topics can be considered. This book also can be employed as one of the books that would be used in general applications programming courses, courses on spreadsheets, or in a computer graphics course.

Educators will also find the book to be a valuable supplement for providing models and ideas that can be used in a variety of college-level mathematically oriented classes such as those in pre-calculus, calculus, linear algebra, geometry, complex analysis, science, and engineering. Because most students in these areas will either already know spreadsheet usage or be able to pick it up rapidly, the book can be used to provide assignments and class projects. Furthermore, it can be used by teachers as a source of ideas and examples for designing illustrations and models for interactive classroom demonstrations. In these demonstrations a teacher or student can lead the class in "What if...?" explorations. This is particularly effective when the curves are shown through liquid crystal display or other projection devices. One can create figures just as they appear in other books and texts, but with the added interactive features. In many of these applications, the book can also be used at the secondary level.

Laboratory Reference. This book also should prove to be a valuable reference book when made available to students in computer, mathematics, and science laboratories, as well as in reference rooms and in libraries.

Personal Reading and Study. Finally, the topic of mathematical curves and how to create them is in itself fascinating and absorbing. Thus, this book with its spreadsheet approach to curve sketching presents a convenient way for individuals to become involved with the topic of curves while using their own computers and spreadsheet programs. The ideas presented should expand a reader's insight into the topics of curves and constructions, and open new avenues for graphical spreadsheet investigations.

TOPICS AND ORGANIZATION

Chapter 1 contains a general introduction to overall spreadsheet operation. This introduction is designed to provide novices with a sufficient background for using spreadsheets effectively in creating graphs of curves. This chapter also contains a survey of the mathematical concepts that are needed in the remainder of the book, and describes how to use the XY graph type in creating graphs of mathematical curves.

Chapter 2 presents a fundamental model for sketching curves that are defined by parametric or polar equations. Modifications of this model are used throughout the book to generate a large number of classical curves. These modifications are implemented in a way that also allows a user to generate a wide range of other extraordinary curves as well. Most sections of the chapter present one specific curve in some detail and contain one or more additional special constructions to generate that curve, usually through families of lines or other curves.

Chapter 3 contains models for constructing such general families of curves as strophoids, pedals, evolutes, and parallel curves. Frequently these curves are formed as envelopes of families of lines, circles, or other curves. Although each model is implemented with one particular base curve, each is designed so that users can experiment with a wide range of underlying curves to produce a great variety of output.

Chapter 4 uses circular spreadsheet references to create an animation effect on a spreadsheet. With each recalculation of the spreadsheet some action takes place. For example, a curve changes, a circle rolls, a construction triangle moves, or curves are traced. These models are particularly effective when implemented on those spreadsheets that allow both the spreadsheet model and a graph to be displayed on the screen simultaneously. However, even in spreadsheets without such a capability, the models are both interesting and instructive for showing how changes in parameters affect the curve that is produced and for illustrating the steps in constructions.

Chapter 5 shows how a spreadsheet can be used to investigate curves that arise in various contexts in the field of complex analysis. In particular, different representations of functions of a complex variable are created through the use of curves and grids. Some illustrative geometrical applications of complex analysis to other fields are initiated as well.

Appendices provide a listing of many of the primary curves and their derivatives for use in modifying the book's examples, an introduction to methods for embedding spreadsheet graphical output in word processing documents, and a guide for using an accompanying disk of spreadsheet examples.

References supply additional sources for descriptions of other curves to study, implement, and investigate on a spreadsheet.

COMPUTER AND SPREADSHEET REQUIREMENTS

This book can be used with an extensive range of the most popular spreadsheets, including *Quattro Pro*, *Lotus 1-2-3*, and *Microsoft Excel*. An accompanying 3½ inch disk in the double density format provides 25 selected examples written in *Quattro Pro 2.0*, *Lotus 1-2-3 2.3*, and *Microsoft Excel 4.0*. The examples on the disk can be run on those or higher numbered versions of the spreadsheets, as well as on *Lotus 1-2-3 2.2*. The specific examples that are contained on the disk are listed in Appendix 3, along with directions for using the disk. In addition, brief instructions are contained on the disk as ACSII text files that can be read by using the Type command of DOS (e.g., Type READ_ME.DOC).

The disk is designed for use with *IBM PC* compatible computers. The files for each of the spreadsheets occupies less than 250 kb. of the disk. To be able to use the disk with Version 2.0 of *Quattro Pro* or Versions 2.3 and 2.2 of *Lotus 1-2-3*, the minimal computer requirements include:

- An *IBM PC* or fully compatible computer using DOS Version 3.3 or later

- A 286 or higher numbered processor

- At least 640 kb. of memory

- A 3½ inch disk drive and a hard drive with at least 1 mb. of available capacity

- A VGA graphics card and color monitor

Additional requirements will be determined by the specific spreadsheet being used. For example, while *Quattro Pro 2.0*, *Lotus 1-2-3 2.3*, or *Lotus 1-2-3 2.2* can use computers with 286 processors, 386 processors are needed for higher numbered versions, with 486 processors recommended for use with *Windows* versions. The *Quattro Pro* and *Lotus 1-2-3* examples do not require the use of *Windows* versions of those spreadsheets.

The specific memory capacity that is required depends upon the particular version of the spreadsheet that is being used. The *Quattro Pro* and *Lotus 1-2-3* files can be run on the versions mentioned above using computers with 640 kb. of memory. However, additional memory is required by higher numbered versions of those spreadsheets, and it can increase the speed and efficiency of operation significantly in any case. The *Microsoft Excel 4.0* examples run in the *Windows* environment, with at least 4 mb. of memory recommended. Specific system requirements of a spreadsheet should be consulted in the spreadsheet user's manual.

The examples that are furnished have been created in an abbreviated form so that they will fit on a small disk. The examples can be expanded either manually through spreadsheet commands or automatically through built-in spreadsheet macros that have been provided. While the examples can be loaded into memory directly from the disk and then expanded, there is not enough room on the disk to store the expanded versions. It is more efficient to first transfer the files to a hard disk, and then save the expanded versions either on a hard disk or on another diskette.

The examples in this book may be implemented on many other spreadsheets as well, as long as the spreadsheet provides the XY graph type and has a *pen-up* capability that can be used to produce breaks in drawing curves. In addition, the computer employed must be capable of supporting one of these spreadsheet programs. If a spreadsheet is designed to run in a *Windows* environment, then that program must be provided as well.

This book has been written in Version 2.0 of *Quattro Pro*. This older version was specifically chosen so that the examples might be implemented even on older, smaller spreadsheets, and so that illustrative examples are as general as possible. While some of the advanced features of the more recent spreadsheet versions can provide for embellishments of some of the models, these newer, larger programs generally are not essential for the examples in this book.

To use the larger spreadsheet models in the book most effectively, it is very helpful to use a computer that has a hard drive, an adequate amount of memory, a VGA graphics card, and a color monitor. Typically, at least 4 mb. of memory are desirable for setups that use current operating systems, the latest spreadsheets, and *Windows* software. However, with some spreadsheets a smaller configuration still may prove to be sufficient. The *IBM PC* compatible computers with faster 386 or 486 processors are superior, especially in running those models that are larger or employ animation. The 486DX processors, with their built-in mathematical co-processor, clearly are more effective, especially with the animated examples. However, even computers with 286 processors and 640 kb. of memory can be used for all of the book's examples when they are constructed in *Quattro Pro 2.0* or *Lotus 1-2-3 2.3*, for example, although user interaction and display updating are carried out more slowly. Similar statements apply to *Macintosh* computers with equivalent *Motorola* processors.

ACKNOWLEDGEMENTS

I wish to express my appreciation for the contributions of several people and institutions. A 1989 Fulbright professorship at The University of Papua New Guinea provided an excellent teaching opportunity that enabled me first to become interested in the graphical aspects of mathematical modeling on spreadsheets. At Whitworth College, my colleagues Randy Michaelis and Ken Pecka provided both encouragement and conceptual suggestions, while Elizabeth Carras and Rick Brodrick contributed technical support. Professor Ross Cutter has furnished invaluable insights in many realms. In addition, Whitworth College's "Mystery Man" supplied valuable funding for travel to professional meetings.

Spreadsheet colleagues Dr. Robert S. Smith of Miami University and Dr. Stephen Comer of The Citadel examined the book and provided helpful comments, as did Kristine Numrich, a Miami University graduate student. Extensive discussions over several years with Dr. Erich Neuwirth of the University of Vienna on all aspects of spreadsheets have been most beneficial, and Professor Richard Guy of the University of Calgary has suggested good ideas for investigating curves. In addition, I wish to thank my wife Susan Arganbright of Gonzaga University for contributing artistic and investigative ideas, and for furnishing insight into various advanced features of Word Perfect.

Finally, I wish to thank all of the people at CRC Press who helped so much to make this book possible. Special gratitude is extended to Wayne Yuhasz, the executive editor, for heading the project, for making valuable suggestions, and for providing encouragement. In addition, valuable editorial assistance and contributions were made by Fern Laplanch, Bonnie Rhodes, and Sarah Fortener.

The manuscript has been prepared in a camera-ready format in *Word Perfect* using a *PostScript* Times Roman 10-point font and a *Hewlett Packard HP Laserjet IIIp* printer. The graphs have been created using *Quattro Pro*.

THE AUTHOR

Dr. Deane E. Arganbright is Professor of Mathematics and Computer Science at Whitworth College in Spokane, Washington. He received B.S. and B.S.Ed. degrees from Bowling Green State University, and M.A. and Ph.D. degrees from the University of Washington. He has also taught at Iowa State University, the University of Papua New Guinea, and Bendigo College of Advanced Education (Australia). In 1989, he returned to the University of Papua New Guinea as a Fulbright Professor.

Professor Arganbright is a pioneer in the development of mathematical applications of spreadsheets. He has written another book on the mathematical applications of spreadsheets and has published many additional articles on this topic in journals and books. In addition, he has given numerous talks at national and regional meetings of mathematics and computer science societies, as well as presentations in China and in Europe. He continues to be interested in developing a variety of approaches for uniting mathematics, mathematical modeling, and mathematical applications through the use of computers. His current emphasis centers on the visualization of mathematics and statistics.

Dr. Arganbright is currently the president of the Inland Northwest Chapter of the Association for Computing Machinery and is a member of the Mathematical Association of America's Committee on Visiting Lecturers. He is a member of the American Mathematical Society, the National Council of Teachers of Mathematics, the Society for Industrial and Applied Mathematics, and the American Statistical Association.

To my wife Susan with love and appreciation

CONTENTS

CHAPTER 1 SPREADSHEET TUTORIAL 1

    1.1   Spreadsheet Operation 1
    1.2   Background Mathematics for Curve Sketching 8
    1.3   Basic Curve Sketching Techniques 12
    1.4   Multiple Curves 17
    1.5   Auxiliary Curves 18
    1.6   Producing Multiple Lines 21
    1.7   Modifications for *Lotus 1-2-3* 24
    1.8   Modifications for *Microsoft Excel* 26

CHAPTER 2 CLASSICAL CURVES 29

    2.1   Parametric Equations 30
    2.2   Primary Parametric Model 32
    2.3   Polar Graphs 36
    2.4   Cardioid 38
    2.5   Limaçon 42
    2.6   Astroid 45
    2.7   Deltoid 52
    2.8   Nephroid 54
    2.9   Lemniscate 57
    2.10   Conic Sections 61
    2.11   Lissajous Figures 63
    2.12   Spirals 65
    2.13   Cycloids and Related Curves 67
    2.14   A Mélange of Curves 73
    2.15   Sketching With Spreadsheet Solver Commands 75

CHAPTER 3 CONSTRUCTIONS 85

    3.1   Strophoids 85
    3.2   Conchoids 89
    3.3   Cissoids 92
    3.4   Pedal Curves 95
    3.5   Negative Pedals 98
    3.6   Inverse Curves 100
    3.7   Parallel Curves 103
    3.8   Evolutes 106
    3.9   Caustics 109
    3.10   An Approximation Model 113

CHAPTER 4 ANIMATION 117

    4.1   Tracing a Curve 117
    4.2   Varying Parameters 119
    4.3   Conic Sections via Circles 122
    4.4   Tracing Conic Sections 127
    4.5   Cycloids as Rolling Circles 132
    4.6   Hypocycloids and Epicycloids 135
    4.7   Involute of a Circle 139
    4.8   Glissettes 142
    4.9   Straight-Edge and Triangle Constructions 144
    4.10   Linkages 146
    4.11   Circular Functions 149
    4.12   Curves of Constant Width 151

CHAPTER 5 CURVES FROM COMPLEX VALUED FUNCTIONS 155

    5.1   Complex Powers 155
    5.2   Complex Roots 159
    5.3   Graphing Complex Functions 161
    5.4   Additional Complex Functions 164
    5.5   Grid Representation of Functions 169
    5.6   Streamlines 173
    5.7   Joukowski Airfoil 176
    5.8   Ovals of Cassini 178
    5.9   Inverses of Complex Functions via Spreadsheet Solvers 179

APPENDICES 183

    A.1   Selected Curves and Their Derivatives 183
    A.2   Graphs in Word Processing 185
    A.3   Disk of Examples 189

REFERENCES 193

INDEX 195

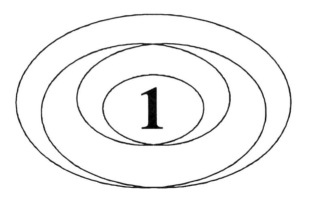

Chapter 1

SPREADSHEET TUTORIAL

The topics presented in this chapter form a basis for creating the spreadsheet models and examples discussed in the remainder of the book. The first section presents an overview of the fundamentals of spreadsheet operation. This material will provide a review for experienced spreadsheet users, and should furnish a good introduction for those to whom spreadsheet usage is new. The second section contains a similar summary of the mathematics employed in designing the book's models.

The remainder of the chapter is devoted to a discussion of various facets of spreadsheet graphing. All of the graphs in this book are created by using the XY graph type, and the discussion centers on how this type can be used for sketching mathematical curves. Many of the procedures and ideas presented in this book represent creative and innovative applications for spreadsheet graphs. These techniques include showing within the same graph curves defined for differing domains, incorporating auxiliary lines, and drawing an entire family of lines or other curves in a single series.

The techniques introduced in this chapter are used throughout the book to create eye-catching curves, to design effective illustrations, to incorporate construction details in a drawing, and to form families of curves that define other curves.

1.1. SPREADSHEET OPERATION

The basic techniques of spreadsheet operation are easy to learn. This section presents a brief overview of those fundamental spreadsheet concepts that are needed to implement the examples in this book. Although specific spreadsheet features vary among the different programs and versions, most are quite similar. Throughout this book the conventions of *Quattro Pro* are used to illustrate the concepts. Sections 1.7 and 1.8 discuss some of the modifications needed for two other leading spreadsheets, *Lotus 1-2-3* and *Microsoft Excel*. Additional details of spreadsheet operation are contained in numerous books and manuals.

The spreadsheet screen format consists of a large array, or table, of rows and columns, much like an accountant's spreadsheet. Rows are identified by positive integers, and columns by letters. An individual array entry, or cell, is identified by row and column, such as C2.

One cell at a time is highlighted on the screen by a cursor. The cursor can be moved throughout the screen by the arrow keys, by using a mouse, or through a GOTO command. Figure 1.1 illustrates the spreadsheet format with the cursor placed at Cell C2. A spreadsheet contains considerably more rows and columns than are displayed at one time.

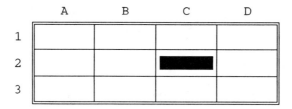

Figure 1.1. Spreadsheet Display.

Into the cell indicated by the cursor one can enter a string (or label), a number, or a formula that references other cells as variables in the formula. The spreadsheet program evaluates each formula by using the values of the cells to which they refer, and displays the resulting values on the screen. If the value of any cell is changed, the spreadsheet is recalculated and the screen display is updated. This provides the spreadsheet's "What if...?" modeling capability.

1.1.A. An Illustrative Example

An illustrative example is presented in Figures 1.2 and 1.3. This model computes the growth of a one-time deposit P in a savings account that pays interest compounded annually at the rate r for n years, $0 \le n \le 66$. The screen output is displayed in Figure 1.2, and the formulas are shown in Figure 1.3. The symbol : in the row identifiers along the left side of each of the displays indicates that a number of rows have been omitted.

```
         A            B
 1    Rate:          5.0%
 2 Principal:    $1,000.00
 3    Year:             3
 4    Balance=   $1,157.63
 5       Year     Balance
 6          0   $1,000.00
 7          1   $1,050.00
 8          2   $1,102.50
 9          3   $1,157.63
10          4   $1,215.51
 :
71         65  $23,839.90
72         66  $25,031.90
```

```
         A              B
 1    Rate:           0.05
 2 Principal:         1000
 3    Year:              3
 4    Balance= @VLOOKUP(B3,T,1)
 5       Year       Balance
 6          0           +B2
 7       1+A6    (1+B$1)*B6
 8       1+A7    (1+B$1)*B7
 9       1+A8    (1+B$1)*B8
10       1+A9    (1+B$1)*B9
 :
71      1+A70   (1+B$1)*B70
72      1+A71   (1+B$1)*B71
```

Figure 1.2. Interest Model Output.

Figure 1.3. Interest Model Formulas.

To create the model, first the indicated descriptive labels are entered in Cells A1..A5 and

B5. Next, the annual rate r is entered as a decimal in Cell B1, and the principal P is set in Cell B2. In this example, $r = 0.05$ (for 5%) and $P = 1000$ (for $1000). The model is arranged so that these parameters of the model are entered at the top of the spreadsheet. To see the effects of changes in the parameters, a user simply needs to change the values of these cells.

The model calculates yearly balances by an elementary recurrence relation. If the balance of the account after n years is X_n, then its value X_{n+1} after 1 more year is obtained by adding the n-th year's interest of rX_n to the previous balance X_n, to give

$$X_{n+1} = X_n + rX_n = (1+r)X_n$$

This expression is implemented in Rows 6..72. Column A provides a count for the years, while Column B calculates the balances for successive years. The initial setting is formed by entering 0 in Cell A6, and reproducing the principal in Cell B6 by entering the formula +B2. The use of a + sign as the first symbol of an expression indicates to the program that the entry is a mathematical expression, or formula, that yields a numerical value. In *Quattro Pro*, expressions that start with +, (, @, or a number are considered to be value expressions. Others are regarded as strings.

Next, to calculate the values at the end of the first year, the formula 1+A6 is entered in Cell A7, generating the number 1. Cell B7 computes the balance after 1 year by the formula (1+B$1)*B6. This can be interpreted as "(1+rate) times the cell above". The use of the $ sign is explained below.

Now the spreadsheet's Copy command can be used to complete the table. To invoke the command menu the symbol / is typed. A user then selects options from the menu that appears, following the prompts that are provided. These also may be selected by using a mouse. The selections below illustrate the copy procedure. The options that a user enters are shown in bold. Frequently choices are made by entering a single letter. In this example, after / is entered, the Edit option is chosen by **E**, followed by **C** for Copy. At this time the menu prompts a user for the cells that are to serve as source and destination ranges. These may be entered either by typing the ranges shown, or by using the arrow keys or a mouse.

> **/** E(dit) C(opy) Source: **A7..B7** Destination: **A8..A72**

The effect of these choices is to copy the formulas in the source block A7..B7 into the destination block A8..B72. Only the first column of the destination needs to be indicated. However, for completeness, each of this book's models lists the entire destination range. In many spreadsheets the more common commands also have shortened forms, such as <CTRL>-C for Copy in *Quattro Pro*.

Before a formula is copied, each of its cell references is designated as being either an absolute or a relative location through the use of the $ symbol. For example, in copying the formula in Cell A7, the location A6 is a relative reference, or "1 + the cell above", because no $ is used. That cell reference changes when the formula is copied. The $ sign designates an absolute reference. When the formula in Cell B7 is copied, the cell reference for the rate should remain constant. Hence, B$1 is written for the rate, indicating an absolute row reference that does not change as the formula is copied down Column B. Again, the previous balance (B6) is relative, so the $ is not used with that reference.

This model also illustrates the use of another spreadsheet feature, the built-in (or library) function. In most spreadsheets, these functions are designated by an initial @. The vertical table lookup function @VLOOKUP is used frequently in this book. In this example the Name command is used first to name the block of cells A6..B72 as T,

> **/ E(dit) N(ames) C(reate) Name: T Block: A6..B72**

Now suppose that a user desires to know the balance of the account at the end of year $N = 3$. The number 3 is entered in Cell B3, and the formula @VLOOKUP(B3,T,1) is placed in Cell B4. This function searches vertically in the initial column of the table T for the occurrence of the value of B3. When it finds the value (here, in Cell A9), it returns the value of the cell that is located one column to the right (or $1157.63). Thus, the last number in the formula provides the amount of column offset. In *Quattro Pro* and *Lotus 1-2-3* the left-most column is counted as the 0-th column, while some spreadsheets, such as *Excel*, start this count at 1. The entries in the initial column of a vertical lookup table must be in ascending order.

It is wise to save a model frequently using the Save command:

> **/ F(ile) S(ave) Name: GROW**

This command saves the program on the current default drive as a file with an extension provided by the program. In *Quattro Pro* the file is saved as GROW.WQ1. To retrieve the file in a subsequent session, the Retrieve command is issued:

> **/ F(ile) R(etrieve) Name: GROW**

1.1.B. Introduction to Graphing

The following description provides a brief illustration of how to create a spreadsheet graph. Additional graphing techniques are discussed in Section 1.3. Here, the spreadsheet's XY graph type is used to create a graph that shows the growth of the savings account of the previous example. The Graph command is employed to select the graph type, to designate the ranges of the *x*- and *y*-values, and to give the graph a name. In this case, the *x*- and *y*-series of the graph consist of the ranges A6..A72 and B6..B72, respectively.

> **/ G(raph) G(raph Type) Type: XY**
> **S(eries) X: A6..A72 1(st y): B6..B72 Q(uit)**
> **N(ame) C(reate) Name: GROWTH Q(uit)**

To view the current graph, a user either chooses the View option while the Graph menu is active, or presses the **<F10>** key at any time. The graph produced for this model is shown in Figure 1.4. Some identifying labels have been included in the graph's construction by using Graph Text command options.

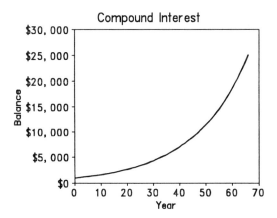

Figure 1.4. Compound Interest.

This example can be concluded by saving the changes that have been made since the previous Save command was issued. Because the file currently has a name (GROW), the spreadsheet supplies that name, although a user has the option of changing it via the Save As command.

> / **F**(ile) **S**(ave) File exists **R**(eplace)

Other spreadsheets have a few different conventions that should be mentioned. In some, such as *Excel*, formulas are identified through an initial = rather than +, library functions do not start with @, graphs may be created in separate files, and the Fill command is used in place of Copy. These and other changes can be learned from manuals and reference books. Some of these are discussed further in Sections 1.7 and 1.8.

1.1.C. Inserting and Deleting

Because some curves require more points to be plotted than do others, it is important to be able to change an existing model to produce either more or fewer points. This is done through the Insert and Delete commands.

First, the number of points can be increased either to extend a curve or to produce one that is smoother. This process is illustrated through the compound interest example of Figure 1.3. Suppose that it is desired to extend the model through the 100th year. To do so, the cursor is moved to Cell A72. The Insert command is then used together with the down arrow key or the mouse to insert 34 blank rows between the current Rows 71 and 72. The new final row is Row 106. Next, the formulas in Row 71 are copied into Rows 72..106. By inserting the rows within the existing block, the model is extended, and the graph is adjusted automatically. Throughout this book this process is referred to as the insert and copy procedure.

> / **E**(dit) **I**(nsert) **R**(ows) Enter Block: **A72..A105**
> / **E**(dit) **C**(opy) From: **A71..B71** To: **A72..A106**

Second, it is possible to delete rows and thereby decrease the size of a model. The process is illustrated by reducing the number of years to 30 in Figure 1.3. Again, the rows to be deleted should lie within the interior of the graph block so that the graph will be adjusted automatically. The cursor is moved to Row 36 (the 30th year), and Rows 36 through 71 are deleted. Next, Row 35 is copied into the new Row 36.

```
/ E(dit)  D(elete)  R(ows)  Enter Block: A36..A71
/ E(dit)  C(opy)  From: A35..B35  To: A36
```

While modifying a model, it may be convenient to turn off a spreadsheet's automatic recalculation feature while changes are being made in order to avoid unnecessary recalculations, and thereby accelerate the modification process. This can be done as follows:

```
/ O(ptions)  R(ecalculation)  M(ode)  M(anual)
  Q(uit)  Q(uit)
```

When this option has been selected, the spreadsheet will be recalculated only if the recalculation key <F9> is pressed. After the changes have been made in the model, the automatic recalculation option can be reactivated by resetting the mode to automatic.

1.1.D. Circular References

If a formula in a given cell references itself, either directly or through a series of other cells, a *circular reference* results. Circular references are employed extensively in Chapter 4 in creating animated models. Here, a brief example is presented to illustrate the use of circular references in finding the sum of the first n integers, $S(n) = 1+2+...+n$. The formulas for the model are given in Figure 1.5, with the output described in Figures 1.6 to 1.8. It may be advantageous to choose row-wise calculation in this example.

```
         A                    B
1  Start>               0
2  N =                  @IF(B1=0,0,1+B2)
3  n                    S(n)
4  @IF(B1=0,0,A7)       @IF(B1=0,0,B7)
5  1+A4                 +A5+B4
6  1+A5                 +A6+B5
7  1+A6                 +A7+B6
```

Figure 1.5. Formulas.

Cell B2 contains the library @IF function. An expression of the form @IF(a,b,c) is read "IF a THEN b ELSE c". In this example, @IF is first used to count iterations in Cell B2. There, the formula @IF(B1=0,0,1+B2) means "IF Cell B1 is 0, THEN 0, ELSE add 1 to B2". Thus, Cell B2 references itself directly.

```
      A    B                    A    B                    A    B
1  Start>  0               1  Start>  1               1  Start>  1
2    N =   0               2    N =   1               2    N =   2
3    n   S(n)              3    n   S(n)              3    n   S(n)
4    0    0                4    3    6                4    6    21
5    1    1                5    4   10                5    7    28
6    2    3                6    5   15                6    8    36
7    3    6                7    6   21                7    9    45
```

Figure 1.6. B1 = 0.   **Figure 1.7.** B1 = 1.   **Figure 1.8.** Recalc.

If Cell B1 is set to 0, then 0 is returned in Cell B2. If any nonzero value (say, 1) is entered in Cell B1, then 1 is added to the previous value of B2 to generate 1. Spreadsheet recalculation can be forced by pressing the <F9> key. Thus, with each recalculation after Cell B2 is set to a nonzero value, 1 is added to Cell B2 to generate successively 1,2,3,4,.... This provides an iteration count, N. In this book such use of Cell B1 is referred to as initializing the model.

Cells A4 and B4 also contain circular references. Column A maintains a point count, n. If Cell B1 is 0, then 0 appears in Cell A4, and the values 1, 2, and 3 are calculated in the subsequent cells of Column A. However, when Cell B2 is set to 1, Cell A4 assumes the previous value of Cell A7 (i.e., 3), and the values 3, 4, 5, 6 are formed in Column A. Thus, the counting is advanced with each recalculation.

Column B determines $S(n)$ as a cumulative sum. If Cell B1 is 0, then Cell B4 is 0, and the remaining values in Column B are computed as the sum of the current n and the cell above, giving 1, 3, 6. When Cell B1 is set to 1, and with each subsequent recalculation, the previous entry in Cell B7 is copied into Cell B4, and the process is repeated. Thus, the output is updated continually with each recalculation.

One potential problem in using circular references is that a spreadsheet must determine which computation to carry out first. Most spreadsheets have built-in conventions for doing this automatically, and the book's graphing examples are designed to fit these. However, sometimes it may be advantageous to specify the recalculation order. To carry out a row by row recalculation, one enters

> **/ O**(ptions) **R**(ecalculate) **O**(rder) **R**(owwise) **Q**(uit) **Q**(uit)

1.1.E. Spreadsheet Model Design

In this book the layouts of the models have been condensed in order to save space. In a well-designed working model, however, it is wise to provide a number of additional features such as

- Comments, instructions, and other documentation for users

- Full headings for input and output cells

- Adequate spacing between the various sections of a model in order to improve the presentation format and readability

8   *Spreadsheet Operation*

In addition, spreadsheets allow the contents of certain cells to be protected from changes. Thus, template models can be constructed in which users can change only the parameters.

Activities

1. Create a spreadsheet model to project the population growth of two nations by modifying the compound interest model and by using the annual growth rate as the interest rate. Almanacs and other library references can serve as data sources for current populations and growth rates. Create an XY graph to illustrate the projected populations.
2. Create models to compute: (a) $n! = (1)(2)(3)...(n-1)(n)$ from $n! = n(n-1)!$; and (b) the Fibonacci numbers, $F(n)$, where $F(1) = F(2) = 1$, and $F(n+2) = F(n+1) + F(n)$, $n > 1$. For these and other mathematical applications, see Arganbright (1985).
3. Create models to compute partial sums of terms of various series, such as $\Sigma(1/n)$ or $\Sigma(1/n^2)$. Next, modify the models to employ circular references that display sets of 20 terms.
4. Create a model for the binomial probability distribution, finding the probability P_i of i successes in n repetitions of an event that has a probability of success p in each trial. Use the recurrence relation $P_0 = q^n$, $P_i = p(n-i+1)P_{i-1}/(qi)$, $i > 0$, where $q = 1-p$. Some applications of spreadsheets for statistics are discussed in Arganbright (1992).

1.2. BACKGROUND MATHEMATICS FOR CURVE SKETCHING

Although the material in this book does not require an extensive amount of advanced mathematics, some mathematical background material may prove helpful. This section provides a very brief overview of several topics that are employed in curve sketching. Additional details may be found in the references, especially in calculus books.

1.2.A. Cartesian Coordinate System

The book's graphs are constructed using the *rectangular*, or *Cartesian*, *coordinate system*. This system is built from two perpendicular lines, the *x*- and *y*-axes, shown in Figure 1.9.

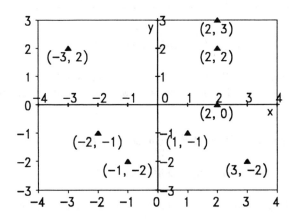

Figure 1.9. Cartesian Coordinates.

The point at which the axes intersect is called the *origin*. Numerical scales are provided for each axis, with positive *x*-values to the right of the origin, and positive *y*-values above the origin. Each point in the plane is determined by its (*x*,*y*)-components relative to these axes. Figure 1.9 illustrates the coordinates of some typical points. The origin is the point (0,0). The axes of the graph have been generated expressly through entries in the model. The spreadsheet *Quattro Pro* provides the *x*- and *y*-coordinate values along the edges.

1.2.B. Vectors in Two Dimensions

A *vector* can be thought of as a directed line segment **v** = [*a*,*b*] that starts from a given point, or *tail*, and extends *a* units in the *x*-direction and *b* units in the *y*-direction. Often the terminal point, or *head*, of a vector is denoted by an arrow. The values *a* and *b* are called the *x*- and *y*-components of the vector. Figure 1.10 illustrates the vector [3,2], with its tail at (2,1) and its head at (5,3). The arrowhead was created by plotting additional short line segments appropriately. Methods for doing this, or using the graph annotator to produce various drawing embellishments, are not discussed here. Throughout the book the coordinates of points are denoted by round brackets and the components of vectors by square brackets.

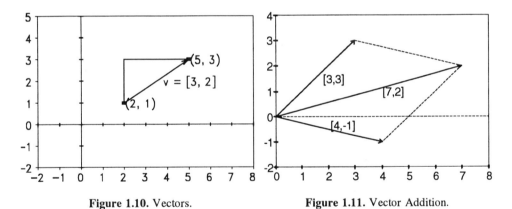

Figure 1.10. Vectors. **Figure 1.11.** Vector Addition.

Where it is convenient, a mixed addition notation is adopted, combining vector and point notation, to generate vector constituents. For example, if the tail of the vector [*a*,*b*] is (x_0,y_0), then its head is at the point (x_1,y_1), where

$$(x_1,y_1) = (x_0,y_0) + [a,b] = (x_0+a, y_0+b)$$

Thus, the vector from (x_0,y_0) to (x_1,y_1) can be written as

$$[a,b] = [x_1-x_0, y_1-y_0]$$

The *length*, *d*, of this vector, which is also the *distance* between the points (x_0,y_0) and (x_1,y_1), is given by

$$d = \sqrt{a^2+b^2} = \sqrt{(x_1-x_0)^2+(y_1-y_0)^2}$$

A vector which has length 1 is called a *unit vector*. If P and Q are points, the distance from P to Q is often denoted by PQ.

The *scalar multiple* of a vector $[a,b]$ by a real number, k, is defined by $k[a,b] = [ka,kb]$. For $k > 0$, the vector $k[a,b]$ is k times as long as $[a,b]$, and it extends in the same direction as $[a,b]$. Thus, if d is the length of $[a,b]$, the vector

$$\mathbf{u} = (1/d)[a,b] = (1/d)[x_1-x_0, y_1-y_0]$$

is a unit vector in the direction of $[a,b]$. For $k > 0$,

$$\mathbf{v} = (k/d)[a,b] = (k/d)[x_1-x_0, y_1-y_0]$$

is a vector of length k in the direction of $[a,b]$. If $k < 0$, the direction is reversed.

In particular, the head (x_2, y_2) of the vector of length k that starts at (x_0, y_0) and goes in the direction from (x_0, y_0) to (x_1, y_1) is

$$(x_2, y_2) = (x_0, y_0) + (k/d)[x_1-x_0, y_1-y_0]$$

Thus, x_2 and y_2 are calculated by

$$x_2 = x_0 + (k/d)(x_1-x_0), \quad y_2 = y_0 + (k/d)(y_1-y_0)$$

The *dot product* of two vectors $\mathbf{u} = [a,b]$ and $\mathbf{v} = [c,d]$ is defined by $\mathbf{u} \cdot \mathbf{v} = ac + bd$. From this it follows that the length of a vector $\mathbf{u} = [a,b]$ can be determined by

$$\sqrt{\mathbf{u} \cdot \mathbf{u}} = \sqrt{a^2+b^2}$$

In many spreadsheets the dot product can be implemented through the library function @SUMPRODUCT. Two vectors $\mathbf{u}$ and $\mathbf{v}$ are *perpendicular*, or *normal*, if $\mathbf{u} \cdot \mathbf{v} = 0$. Thus, the vectors $[x,y]$ and $[-y,x]$ are perpendicular.

Because the slope of the vector $[a,b]$ is $m = b/a$ for a nonvertical vector, the angle that the vector makes in reference to the positive x-axis is $\alpha = \arctan(b/a)$. Finally, the *sum* of two vectors, illustrated in Figure 1.11, is defined as

$$[a,b] + [c,d] = [a+c, b+d]$$

1.2.C. Derivatives, Slopes, Tangents, and Normals

Some geometric constructions use lines that are either tangent or normal to a given curve. From calculus, if a curve is given by $y = f(x)$, then the slope m of the tangent line at the point $(x_0, f(x_0))$ on the curve is the derivative

$$m = dy/dx = f'(x_0)$$

Thus, an equation of the tangent line at that point is

$$y = f(x_0) + f'(x_0)(x-x_0)$$

A line is normal to the curve of $y = f(x)$ at a point $(x_0, f(x_0))$ if it is normal, or perpendicular, to the tangent line at that point. If m is the slope of the tangent line, then the slope of the normal line is $-1/m$, unless $m = 0$. Thus, for $m \neq 0$, the equation of the normal line at the point $(x_0, f(x_0))$ is

$$y = f(x_0) - (x-x_0)/f'(x_0)$$

When $m = 0$, the normal line is vertical, $y = x_0$. Tangent and normal lines are illustrated in Figure 1.22 of Section 1.5. If x and y are defined in terms of a parameter θ, with $y = g(\theta)$, $x = h(\theta)$, then from calculus

$$dy/dx = (dy/d\theta)/(dx/d\theta) = g'(\theta)/h'(\theta)$$

while if a curve is defined in polar coordinates as $r = f(\theta)$, $x = r\cos(\theta)$, $y = r\sin(\theta)$, then

$$dy/dx = [(dr/d\theta)\sin(\theta)+r\cos(\theta)]/[(dr/d\theta)\cos(\theta)-r\sin(\theta)]$$

Because not all spreadsheets allow the use of non-ASCII characters such as θ, it may be convenient to use t or another letter as a parameter, especially for nonpolar coordinates.

1.2.D. Trigonometry

Using the triangle shown in Figure 1.12, with t replacing θ, for an angle θ the basic trigonometric relations are given by

$$\cos(\theta) = x/r, \ \sin(\theta) = y/r, \ \tan(\theta) = \sin(\theta)/\cos(\theta) = y/x$$

so that $x = r\cos(\theta)$ and $y = r\sin(\theta)$. If $r = 1$, then $x^2 + y^2 = 1$, (x,y) lies on a unit circle, and $\cos^2(\theta) + \sin^2(\theta) = 1$. Thus, $[\cos(\theta),\sin(\theta)]$ is a unit vector. Also, the head (x_1,y_1) of a vector of length k in the direction θ that starts at a point (x_0,y_0) is

$$(x_1,y_1) = (x_0,y_0) + k[\cos(\theta),\sin(\theta)]$$

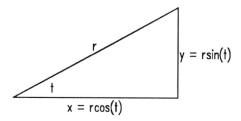

Figure 1.12. Trigonometric Relations.

Three other basic trigonometric functions are defined by

$$\sec(\theta) = 1/\cos(\theta), \ \cot(\theta) = 1/\tan(\theta), \ \csc(\theta) = 1/\sin(\theta)$$

Other primary identities include:

$$\cos^2(\theta) + \sin^2(\theta) = 1, \ 1 + \tan^2(\theta) = \sec^2(\theta), \ \sec^2(\theta) - 1 = \tan^2(\theta)$$

Throughout the book, generally θ is used as the variable for trigonometric functions. However, because most spreadsheets do not support θ as a graph symbol, t is used in some graph displays.

1.2.E. Miscellaneous Topics

The library modulo function, @MOD(n,m), returns the integer remainder that results when the integer n is divided by the integer m. For example, @MOD(23,5) = 3, @MOD(21,3) = 0, and @MOD(13,6) = 1.

If each member of a family of curves is tangent to a curve C, then C is called an *envelope* of the family. As an example, Figure 2.46 in Chapter 2 illustrates a deltoid as an envelope of lines.

Activities

1. Use a spreadsheet to produce Figure 1.10. Then modify the model to compute and graph the sum of two vectors, as in Figure 1.11.
2. Create spreadsheet graphs of $y = \sin(x)$ and $y = \cos(x)$ from $x = -\pi$ to $x = \pi$. Create graphs of additional trigonometric functions as well.
3. Illustrate $\cos^2(\theta) + \sin^2(\theta) = 1$ by computing @SIN(B1)^2 + @COS(B1)^2 for various values of θ placed in Cell B1.
4. Use the spreadsheet @SUMPRODUCT function to compute dot products of vectors in $\mathbf{R}^n$ for $n = 2,3,\ldots$ The set $\mathbf{R}^n$ consists of n-dimensional vectors $[x_1,x_2,\ldots,x_n]$. The dot product in $\mathbf{R}^n$ is defined by $[x_1,x_2,\ldots,x_n] \cdot [y_1,y_2,\ldots,y_n] = x_1y_1 + x_2y_2 +\ldots+ x_ny_n$.
5. Using definitions from calculus, compute the components of the projection of a vector **v** onto a vector **u** in $\mathbf{R}^2$.

1.3. BASIC CURVE SKETCHING TECHNIQUES

Creating a graph for a curve using a spreadsheet is a straightforward process. Basically, the spreadsheet draws a curve by plotting a large number of points (x,y) and connecting successive points by line segments. If the consecutive points are reasonably close to each other, the resulting graph has the appearance of a smooth curve.

1.3.A. Basic Example

As a first example, the basic graphing technique is illustrated by producing a graph for the curve defined by $y = ax^2$, $-1 \leq x \leq 1$. The model is described in Figures 1.13 and 1.14, with the screen output shown on the left and the formulas listed on the right. In this example, as in many of this book's models, Column A is left blank for other possible uses, such as supplying headings for sections of a model.

```
        B         C       D       E
1  Curve: y = ax²  Title:
2  Init x0:   -1.00   x  =   1.1
3  Step  h:    0.10   y  =   0.5
4     a  =       2
5       x        y1          y2 label
6    -1.00     2.00
7    -0.90     1.62
8    -0.80     1.28
9    -0.70     0.98
:
25    0.90     1.62
26    1.00     2.00
27    1.10                0.50 y=2x^2
```

Figure 1.13. Output.

```
        B             C              D
1  Curve: y = ax²
2  Init x0:     -1
3  Step  h:     0.1
4     a  =       2
5      x         y1                  y2
6    +C2         +C$4*B6*B6
7    +B6+C$3     +C$4*B7*B7
8    +B7+C$3     +C$4*B8*B8
9    +B8+C$3     +C$4*B9*B9
:
25   +B24+C$3    +C$4*B25*B25
26   +B25+C$3    +C$4*B26*B26
27   +E2                            +E3

E27: +"y = "&@STRING(C4,0)&"x^2"
```

Figure 1.14. Formulas.

As in Section 1.1, the parameters of the model are entered at the top of the spreadsheet. The initial x-value, x_0, is entered in Cell C2, with the size h of the step increment between successive x-values set in Cell C3. Here, $x_0 = -1.0$ and $h = 0.1$. The value of the curve parameter a is entered in Cell C4.

Next, the (x,y) coordinates of points on the curve are generated in Columns B and C. First, the initial x-value is reproduced in Cell B6 by +C2. Then the formula +B6+C$3 is entered in Cell B7 to compute the succeeding x-value as the sum of the previous x-value (B6) and the step size h (C$3). Finally, the Copy command is used to copy the formula in Cell B7 into Cells B8..B26.

Again, the effect of the $ sign for absolute references can be observed. When the formula in Cell B7 is copied down Column B, the $ ensures that the 3 in the row reference for the step increment, h, remains constant. On the other hand, in the absence of $, the location B6 is a relative reference. The y-values are created similarly in Column C by entering +C$4*B6*B6 for ax^2 in Cell C6, and then copying that expression into Cells C7..C26. Notice that alternatively, +C$4*B6^2 can be used for ax^2.

Next, the Graph command is employed to create a graph of the curve. As always in this book, the XY graph type is selected. Cells B6..B26 and C6..C26 are selected for the x-series and the first y-series, respectively. The XY graph type plots points by their (x,y)-coordinates, connecting the points with line segments. However, a user has the option of plotting only line segments, only symbols for the points, both lines and symbols, or neither. Figure 1.15 shows the graph produced by using both points and lines. Similarly, horizontal and vertical grid lines can be displayed or not. Once the graph has been created, a user can display it by using the Graph View command, name and save it as part of the model by using the Graph Name command, and then save the entire model as a spreadsheet file by using the File Save command.

Figure 1.16 shows another graph formed from the same model, but not including the grid lines. Here, the option to plot only the lines has been chosen in order to produce a smooth curve. Many spreadsheets allow text to be displayed within a graph. This graph displays the equation for the curve. The (x,y)-coordinates of the grid location in the graph at which the equation is to appear are entered in Cells E2..E3. Then the formulas +E2 and +E3 are placed in Cells B27 and D27, the x-axis series is extended through Row 27, and the second y-series is defined as D6..D27. Neither points nor lines are displayed for the second y-series.

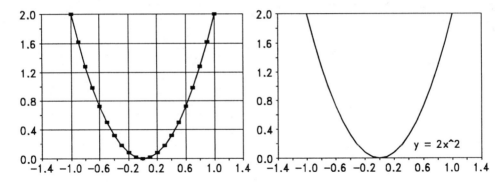

Figure 1.15. Grid with Markers and Lines. **Figure 1.16.** Primary Display Format.

Next, either a string such as "$y = 2x^2$" or the string formula shown is entered in Cell E27. Through the use of a Graph option, Column E provides interior labels for the second series.

> / G(raph) C(ustomize Series) I(nterior Labels) 2nd Series
> Range: **E6..E27** C(enter) Q(uit) Q(uit) Q(uit)

This causes the equation to appear at the chosen point, here, (1.1,0.5). Most of the graphs shown in the book could display equations in this way. However, in order to conserve space, these equations generally are not specified. Providing them is a good exercise. By including string formulas to display an equation in a graph, the equation is updated automatically as parameters are adjusted.

Once the model has been created, the effects produced by changing the parameter a in Cell C4 can be easily investigated. Figure 1.17 shows the graph that is produced by setting $a = 8$, $h = 0.05$, and $x_0 = -0.5$.

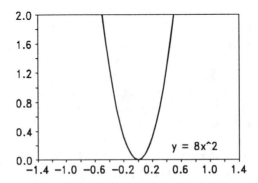

Figure 1.17. $f(x) = 8x^2$.

This book uses the following notation to describe the copy and graph ranges for each of its models. In subsequent examples, a similar table will appear beneath each model's description.

Copy:		Graph:			
From	To	Series	Cells	Labels	Purpose
B7	B8..B26	X	B6..B27		x-axis
C6	C7..C26	1	C6..C26		curve
		2	D6..D27	E6..E27	equation name

1.3.B. Embellishments

Various other graphic options for use in creating graphs that are both appealing and instructive can be investigated by reading manuals and experimenting on the computer. Such options include selecting curve colors and styles, hiding the x- and y-scale markings, hiding the graph's grid, and changing the marker symbols. One of the features used in many of this book's examples is to hide the graph's x- and y-scale markings and axes. The *Quattro Pro* commands that follow accomplish this task, producing the graph of Figure 1.18 from the previous model. The effect is produced by setting the grid and background colors to be identical and choosing the Hide option for the scale formats. Experimentation may be necessary with other spreadsheets.

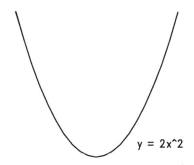

Figure 1.18. Hidden Attributes.

```
/ G(raph)  O(verall)  G(rid)  C(lear)  G(rid Color)  Bright White
           F(ill Color)  Bright White  Q(uit)
  O(verall)  B(ackground Color)  Bright White
  Q(uit)  Q(uit)
  X(-axis)  F(ormat)  H(ide)  Q(uit)
  Y(-axis)  F(ormat)  H(ide)  Q(uit)  Q(uit)
```

1.3.C. Obtaining Correct Proportions

There is one extremely important graphing aspect that should be mastered in order to produce graphs that exhibit a correct perspective, or ratio aspect, between their x- and y-

dimensions. Correct graphs require that identical scales be shown for both the *x*- and *y*-axes. With some spreadsheets the *x*- and *y*-axes are scaled automatically so as to fill the screen. Frequently this means that different scales are adopted for the *x*- and *y*-axes, producing distorted curves. For example, circles may appear as elongated ellipses.

To overcome this difficulty, the highest and lowest values of *x* and *y* can be set manually through a Graph command option. A little experimenting with this procedure is quite helpful. In *Quattro Pro* it is generally best to keep the *x*- and *y*-ranges in an approximate ratio of $x:y = 4:3$. For example, one choice might be $-4 \leq x \leq 4$ and $-3 \leq y \leq 3$. In addition, the use of headings and the font size used for the headings and for the numerical tick marks on the *x*- and *y*-axes will influence the final choice of *x*- and *y*-dimensions. The process used in *Quattro Pro* to create the graph of Figure 1.16 is listed below. The ability to set *x*- and *y*-increments is particularly helpful in refining the resulting proportions.

```
/ G(raph)  X(axis)  S(cale)  M(anual)
  L(ow): -1.4  H(igh): 1.4  I(ncrement): 0.2
  Y(axis)  S(cale)  M(anual)
  L(ow): 0.0  H(igh): 2.0  I(ncrement): 0.2
```

1.3.D. Embedding a Graph

It can be quite effective to embed a graph within the active spreadsheet. When this is done, the effects on the graph are seen immediately as parameters are changed. As discussed in Chapter 4, this format can also be used to effect animation. To insert the current graph into the spreadsheet display, the graphic display mode must first be selected. Then the following commands are issued, with the arrow keys or the mouse used to define the area into which the graph is to be embedded.

```
/ O(ptions)  D(isplay Mode)  B: Graphics Mode  Q(uit)
/ G(raph)  I(nsert)  Enter graph name: <Current>
  Enter area for graph: {Select}
```

Activities

1. Modify the model of Figure 1.14 to create graphs for other curves. For example, produce graphs for each functions $y = x^2 - 0.3x + 2$, $y = \sin(x)$, $y = e^x$, and others.
2. Experiment with setting the *x*- and *y*-scales manually, and obtain graphs that exhibit good perspectives. Hide the axes and scales, and embed the graph in the spreadsheet.
3. Create a model that contains *n* as a parameter, and uses a circular reference to update *n*, $n = 0, 1, 2, \ldots$, with each recalculation. Create a graph of $y = (x-n)^2$ and embed it in the spreadsheet. Observe the resulting movement that is produced as *n* is continually updated through recalculation.

1.4. MULTIPLE CURVES

In order to produce a wide range of graphical displays it is vital to be able to create several curves in the same xy-grid, even though the curves are defined for different sets of values of x. To illustrate one way of doing this, the graphs of a function $y = f(x)$ and its inverse relation that are shown in Figures 1.19 and 1.20 are produced by the model of Figure 1.21. An inverse relation is itself a function only if f is a 1-1 function.

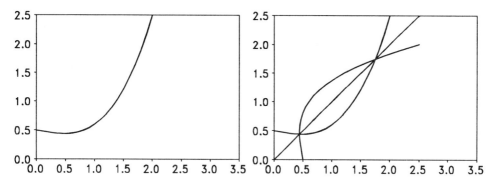

Figure 1.19. Function, $f(x) = 0.3x^3 - 0.2x + 0.5$.   **Figure 1.20.** Function and Inverse.

To create the model, first the initial x-value, $x_0 = 0$, and the step size, $h = 0.1$, are set in Cells D1 and D2. The (x,y)-values for the curve $y = f(x)$ are calculated in Columns B..C of Rows 5..25. The initial x-value is generated first in Cell B5 as +D1. The next x-value, $x+h$, is determined in Cell B6 as +B5+D$2. The latter formula is then copied down Column B through Row 25. In the copying process, x (B5) is a relative reference, while h (D$2) is absolute. Next, functions values $y = 0.3x^3 - 0.2x + 0.5$ are created in Column C by entering the formula 0.3*B5^3−0.2*B5+0.5 in Cell C5 and copying it down Column C through Cell C25. The graph of Figure 1.19 is created by using the ranges B5..B25 and C5..C25 for the x- and first y-series, respectively.

```
         B         C         D         E
1    Initial x =          0.0
2    Step size h=         0.1
3
4        x        y1        y2        y3
5     0.00      0.50                 0.00
6     0.10      0.48                 0.10
7     0.20      0.46                 0.20
  :
24    1.90      2.18                 1.90
25    2.00      2.50                 2.00
26   Inverse:
27    0.50                0.00       0.50
28    0.48                0.10       0.48
29    0.46                0.20       0.46
  :
46    2.18                1.90       2.18
47    2.50                2.00       2.50

B5: +D1      B6: +B5+D$2
C5: 0.3*B5^3-0.2*B5+0.5
E5: +B5
B27: +C5    D27: +B5   E27: +B27
```

Figure 1.21. Function and Inverse.

18 Multiple Curves

Copy:		Graph:			
From	To	Series	Cells	Labels	Purpose
B6	B7..B25	X	B5..47		x-axis
C5..E5	C6..E25	1	C5..25		curve
B27..E27	B28..E47	2	D5..47		inverse
		3	E5..47		y = x

To generate points for the inverse relation, the model produces points of the form (y,x) in a straightforward way in Rows 27..47 from each pair (x,y) of the original curve. First, the initial y-value is reproduced in Cell B27 as +C5, with the initial x-value in Cell D27 as +B5. These formulas are then copied through Row 47.

Next, Graph commands are employed to designate Cells B5..B47 as the x-series, C5..C25 as the first y-series, and D5..D47 as the second y-series. The model also generates the y-values for the line $y = x$ in Column E as the third y-series to illustrate that the other two curves are reflections of each other in that line. The blank cells in the y-series produce no points in the graph. The resulting graph is shown in Figure 1.20.

In some spreadsheets, if the heading "Inverse:" in Figure 1.21 is placed in Cell B26, it may be interpreted as 0 in a graph. Placing it in Cell A26 and significantly reducing the width of Column A allows the heading to be used without affecting the graph.

Activities

1. Use the model of Figure 1.21 to create graphs of various functions and their inverses. Determine which inverses are themselves functions. Insert additional points as necessary for the original and inverse curves. Select colors and styles for different curves to produce good visual effects. Curves to investigate include $f(x) = \sin(x)$, $f(x) = \exp(x)$, and $f(x) = x^2$.
2. When several functions are defined on the same domain, the same x-values can be used to graph different sets of y-values in adjacent columns. Use this technique to plot the graph of $f(x) = \sin(x)$ and the following sequence of Taylor polynomial approximations for f on the interval $-2 \leq x \leq 2$: (a) $f_1(x) = x$; (b) $f_2(x) = x-x^3/6$; (c) $f_3(x) = x-x^3/6+x^5/120$; (d) $f_4(x) = x-x^3/6+x^5/120-x^7/5040$.

1.5. AUXILIARY CURVES

One creative benefit that spreadsheets provide for graphing is the ability to produce a great variety of auxiliary lines. The next model not only draws a curve, but it also creates the tangent and normal lines at any point that is selected on the curve. The graph of the function $y = 0.2x^3-0.5x+1.0$ in Figure 1.22 is created by the model of Figure 1.23. Due to space considerations, in this and in subsequent models, the underlying formulas are listed beneath the output display rather than in the array format.

The curve is created in Columns B..C of Rows 7..27 as in the preceding section. Points on the curve are numbered in Column A by first entering 0 in Cell A7 and 1+A7 in Cell A8. The latter formula is then copied into Cells A9..A27. This numbering technique is used continually throughout the book without further elaboration. Next, the slope of the tangent line is computed at each point. Using calculus, the slope *m* is determined as the derivative by $m = dy/dx = 0.6x^2 - 0.5$. Thus, the formula 0.6*B7^2−0.5 is entered in Cell D7 and then copied down Column D. An XY graph is formed as before, with Columns B and C providing the *x*-series and the first *y*-series, respectively.

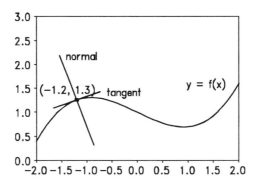

Figure 1.22. Tangent and Normal to Curve.

Now the auxiliary line segments are constructed. First, the Name command is used to designate the range of cells A7..D27 as the table T. The number, *n*, of the point at which the lines are drawn is entered by a user in Cell A3. Here, *n* = 4. Values from table T are next read into Row 3 via table lookup functions.

The formulas in Cells B3..D3 use the value of *n* in Cell A3 to look up the coordinates of (x_0,y_0) and the corresponding derivative m_0 in the table T. For example, Cell B3 locates x_0. The formula @VLOOKUP(A3,T,1) looks vertically down the initial column of T for the occurrence of A3. For *n* = 4 this value is located in Cell A11. The formula returns the entry located 1 column to the right, or -1.2. The formulas for y_0 and m_0 in Cells C3 and D3 similarly return values from the second and third columns to the right. Cell E3 uses the value of m_0 to calculate the slope of the normal line as $m_1 = -1/m_0$.

Rows 28..29 construct the endpoints of a line segment that is tangent to the curve at (x_0,y_0). The length, *k*, for half of the segment is entered in Cell F3. Because the vector $[1,m_0]$ has slope m_0, this vector can be used for the tangent to the curve at the point. Its length is

$$d = \sqrt{1+m_0^2}$$

Using the analysis of Section 1.2, a vector of length *k* that goes in the same direction as $[1,m_0]$ is $(k/d)[1,m_0]$, so the endpoints of the desired tangent segment are

$$(x_0,y_0) \pm (k/d)[1,m_0]$$

As a result, in Cells B28..E28 and B29..E29 the formulas are

$$x_0-k/d, \quad y_0-(k/d)m_0 \quad \text{and} \quad x_0+k/d, \quad y_0+(k/d)m_0$$

The graph's *x*-axis series is extended through Cell B29, and a second *y*-series is defined as E7..E29 to create the desired tangent line segment.

Auxiliary Curves

Columns B and F of Rows 30..31 create the normal line similarly, using a slope of m_1 (Cell E3) and a length k_1 (Cell G3). Again, the x-axis series is extended, and Cells F7..F31 are defined as the third y-series. Finally, the coordinates of the chosen point itself are formed in Cells B32 and G32, and Cells G7..G32 are designated as the fourth y-series, and once again the x-axis is extended. The chosen point is displayed in the symbol format.

Finally, the graph can be enhanced by labeling points and lines via the interior label option of the Graph command. For example, the string "tangent" is placed in Cell H29, with Cells H7..H29 designated as the interior labels for the second y-series. Other labels are implemented similarly, primarily by using descriptive words as the strings. However, Cell H32 uses string functions to display the coordinates of the chosen point as a variable string. The formula is +"("&@STRING(B32,1)&","&@STRING(G32,1)&")". The string function @STRING(B32,1), for example, converts the numerical value of Cell B32 into a string that shows one decimal. The symbol & represents string concatenation, in which two strings are merged to form a single string.

```
         A      B     C       D      E      F       G         H
  1          Select Point:   Slopes:      Lengths:
  2          n     x0    y0     m0     m1   tan:k   nor:k1
  3          4    -1.2  1.254  0.364  -2.75   0.5    1.0
  4
  5                            Slope   Tan   Norm   Point
  6          n     x     y1      m     y2     y3     y4     Label
  7          0    -2.0  0.400  1.900
  8          1    -1.8  0.734  1.444
  9          2    -1.6  0.981  1.036
 10          3    -1.4  1.151  0.676
 11          4    -1.2  1.254  0.364
  :
 26         19    1.8   1.266  1.444
 27         20    2.0   1.600  1.900                         y = f(x)
 28         P0-  -1.67                1.083
 29         P0+  -0.73                1.425                  tangent
 30         Q0-  -1.54                       2.194           normal
 31         Q0+  -0.86                       0.315
 32         P    -1.2                               1.254  (-1.2,1.3)

B3:  @VLOOKUP(A3,T,1)        C3:  @VLOOKUP(A3,T,2)
D3:  @VLOOKUP(A3,T,3)        E3:  -1/D3
A7:  0         A8:  1+A7     B7:  -2        B8:  +B7+0.2
C7:  0.2*B7^3-0.5*B7+1       D7:  0.6*B7^2-0.5
B28: +B3-F3/@SQRT(1+D3^2)    E28: +C3-F3*D3/@SQRT(1+D3^2)
B29: +B3+F3/@SQRT(1+D3^2)    E29: +C3+F3*D3/@SQRT(1+D3^2)
B30: +B3-G3/@SQRT(1+E3^2)    F30: +C3-G3*E3/@SQRT(1+E3^2)
B31: +B3+G3/@SQRT(1+E3^2)    F31: +C3+G3*E3/@SQRT(1+E3^2)
B32: +B3       G32: +C3
H32: +"("&@STRING(B32,1)&","&@STRING(G32,1)&")"
```

Figure 1.23. Tangents and Normals.

Copy:		Graph:			
From	To	Series	Cells	Labels	Purpose
A8..B8	A9..B27	X	B7..B32		x-axis
C7..D7	C8..D27	1	C7..C27	H7..H27	curve
		2	E7..E29	H7..H29	tangent line
		3	F7..F31	H7..H31	normal line
		4	G7..G32	H7..H32	point

Activities

1. Enter other values for n in Cell A3 and observe the lines that are formed.
2. Replace the original function by another one. Remember to change both the function in Column C and its derivative in Column D. Enter the formulas in Cells C6..D6 and copy. Observe the tangent lines that are formed at high and low points on the curve.
3. Amplify the model in Figure 1.16 to show the tangent lines at the corresponding points (a,b) and (b,a) on the original and the inverse curves.
4. Create a graph that contains graphs of both of the curves $y = f(x)$ and $y = f'(x)$. Use, for example, $f(x) = \sin(x)$, $f'(x) = \cos(x)$.

1.6. PRODUCING MULTIPLE LINES

In many later examples, families of large numbers of lines, circles, or other curves, will be constructed. This section describes one scheme for doing this. The model of Figure 1.25 illustrates the process by generating a small family of three regularly spaced lines, each of which is normal to a given base line. The output is shown in Figure 1.24.

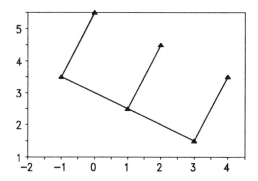

Figure 1.24. Multiple Lines.

The parameters b and m for the base line $y = b + mx$ are set in Cells F2 and H2, with the initial x-coordinate, x_0, entered in Cell D2. Three regularly spaced points on the base line are generated in Rows 4..6. These points are numbered $n = 0, 1, 2$ in Column C. Their x-coordinates are created in Column D as x_0, $x_1 = x_0 + 2$, $x_2 = x_1 + 2$, while their y-values are generated as $b + mx$ in Column E by placing +F$2+H$2*D4 in Cell E4 and copying it.

In Columns G..H of Rows 4..6 the model calculates the coordinates of a second point on the lines that are normal to the base line at each point. Because the slope of the normal line is $-1/m$, the vector $[1,-1/m]$ is normal to the base line. Therefore, each normal line is determined by the given point (x,y) on the base line together with the point

$$(x_a, y_a) = (x,y) + [1,-1/m]$$

Thus, $x_a = x + 1$, and $y_a = y - 1/m$.

To produce these points, first $x_a = 1+x$ is created in Cell G4 as 1+D4. Then $y_a = y-1/m$ is formed in Cell H4 as +E4−1/H$2. Next, the formulas in Cells E4..H4 are copied down through Row 6 and the block C4..H6 named T.

22 Producing Multiple Lines

The remainder of the model produces the required normal lines. This process will be used later to generate many more than three lines. Thus, a technique for drawing the multiple lines that uses only one column for their *y*-values must be devised. Although the XY graph type connects consecutively listed points with line segments, most spreadsheets implement a "pen-up" capability through the "not available" function, @NA. When such a point is encountered in creating a graph, the drawing routine is temporarily discontinued, or "the pen is lifted from the paper". Inserting this function in the *x*- and *y*-series for an XY graph creates a gap to separate the points of distinct lines.

```
            B     C      D      E     F    G     H
         1 Parameters:
         2        x0 =  -1     b =    3   m =  -0.5
         3 Data:   n    x1     y1    y2   xa    ya
         4         0    -1    3.5          0   5.5
         5         1     1    2.5          2   4.5
         6         2     3    1.5          4   3.5
         7 Lines:
         8         n   mod
         9         0    0    -1          3.5
        10         0    1     0          5.5
        11         0    2    NA           NA
        12         1    0     1          2.5
        13         1    1     2          4.5
        14         1    2    NA           NA
        15         2    0     3          1.5
        16         2    1     4          3.5

        D4: +D2    D5: 2+D4   D6: 2+D5
        E4: +F$2+H$2*D4
        G4: 1+D4        H4: +E4-1/H$2
        B9: 0           B10: +B9+(C10=0)
        C9: 0           C10: @MOD(1+C9,3)
        D9: @CHOOSE(C9,@VLOOKUP(B9,$T,1),
                       @VLOOKUP(B9,$T,4),@NA)
        F9: @CHOOSE(C9,@VLOOKUP(B9,$T,2),
                       @VLOOKUP(B9,$T,5),@NA)
```

Figure 1.25. Multiple Lines.

Copy:		Graph:			
From	To	Series	Cells	Labels	Purpose
D5	D6	X	D4..D16		x-axis
E4..H4	E5..H6	1	E4..E6		line
B10..C10	B11..C16	2	F4..F16		normals
D9..F9	D10..F16				

Thus, a formula is designed to produce three points — the two endpoints and the @NA as a separator. The formulas in Column C generate repeated sets of the values 0, 1, 2 to classify these three points. To form these, 0 and @MOD(1+C9,3) are first entered into Cells C9 and C10, respectively. Then the latter formula is copied down Column C to generate a count modulo 3.

Column B numbers these sets of three points to correspond to the points in Rows 4..6. First, 0 is entered in Cell B9 and +B9+(C10=0) in Cell B10. Then the latter formula is copied down Column B. A logical expression such as C10=0 is false (giving a value of 0) unless in fact C10 = 0 (giving a value of 1). Thus, when an entry in Column C is 0 (indicating the start of a new set of 3) the counter in Column B is incremented by 1, otherwise it has the same value as that of the cell above.

Cell D9 uses the @CHOOSE function to generate the 3 types of points. An expression

$$@CHOOSE(C9, Option0, Option1, Option2)$$

returns to the cell containing it, either Option0, Option1, or Option2, depending upon whether the value of Cell C9 is 0, 1, or 2. In Cell D9 of Figure 1.25, when C9 = 0, @VLOOKUP(B9,$T,1) returns the base line's x-value; @VLOOKUP(B9,$T,4) returns the x-value of the other point on the normal line when C9 = 1; and when C9 = 2, @NA is chosen. The formula in Cell F9 serves the same purpose for y-values. Thus, the indicated formulas are entered in Cells D9..F9 and copied through Row 16. The formulas generate points alternately on and off the base line.

The graph is formed by selecting D4..D16 as the x-series, E4..E6 as the first y-series (the base line), and F4..F16 as the second y-series (the normal lines).

Activities

1. Use the insert and copy procedure to create more points for the base line, and to produce more normal lines. For example, the graph of Figure 1.26 can be created by replacing the formula in Cell D5 by 0.2+D4 and copying it into Cell D6. Next, more points are created numbered and numbered 0..20 by inserting new rows following the current Rows 5 and 6 and then copying the formulas from Row 6. The insert and copy procedure is used again between the current Rows 15 and 16.

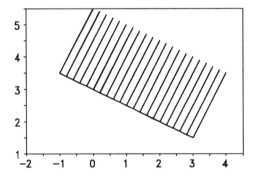

Figure 1.26. Insert and Copy.

2. Rather than creating a family of lines that are parallel to each other, choose a fixed point off a given base line, and create a family of lines that extend from the fixed point to the points on the line as in Figure 1.27.
3. Replace the base line of Figure 1.25 with another curve. Include a computation for the slope at each point. Figure 1.28 was generated from $y = x^2$, $y' = 2x$. What occurs when $x < 0$? To handle this situation, generate normal lines that extend on both sides of the base curve so that the lines all have the same length.

24 Producing Multiple Lines

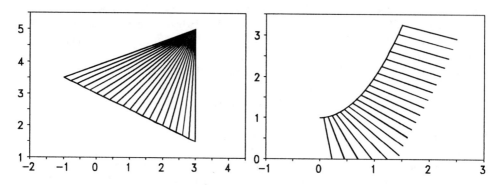

Figure 1.27. Lines from a Point.   **Figure 1.28.** Normals to $y = x^2$.

4. Design a model to generate and plot points of the form $(n,\sin(an))$ for the positive integers $n = 1,2,\ldots,1000$. Create an XY graph that shows neither lines nor symbols. Instead, create a column in which each of the entries is a dot, and use that column to provide interior labels for each point on the graph. Examine the output for several values of the parameter a. Figures 1.29 and 1.30 show the results for $a = 1.0, 0.7$. If computer memory permits it, increase the number of points in order to produce significantly sharper displays. This topic is discussed in Hardin and Strang (1990).

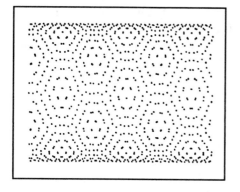   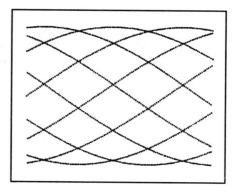

Figure 1.29. $(n,\sin(1.0n))$.   **Figure 1.30.** $(n,\sin(0.7n))$.

1.7. MODIFICATIONS FOR *LOTUS 1-2-3*

Most of the commands in the several versions of *Lotus 1-2-3* are very similar to those of *Quattro Pro*. Spreadsheet manuals and books can provide most of the particulars of these commands, including those on formatting elements of the screen display. This section discusses a few of the features of *Lotus 1-2-3* that are used frequently with the models in this book, especially those related to graphic considerations.

For example, the following commands carry out the standard insert and copy procedure to insert additional rows between Rows 4 and 5 of a given model, and then save the resulting changes.

> / W(orksheet) I(nsert) R(ow) Rows to insert: **A5..A20**
> / C(opy) From: **A4..D4** To: **A5..A20**
> / W(orksheet) S(ave)

- There are some differences between the graph commands of *Lotus 1-2-3* and *Quattro Pro*. Indeed, the graph commands vary somewhat even among the different versions of *Lotus 1-2-3* itself. In *Lotus 1-2-3*, the y-series are designated as A, B, C,..., rather than first, second,..., as in *Quattro Pro*. To illustrate the method of defining a graph in *Lotus 1-2-3*, suppose that a model has been created so that the (x,y) graph coordinates lie in Columns E..F of Rows 6..10. The commands below will create an XY graph.

> / G(raph) T(ype): **XY** X: **E6..E10** A: **F6..F10** O(ptions)
>   G(rid) C(lear)
>   S(cale) Y(-scale) M(anual) L(ower): **-2** U(pper): **2** Q(uit)
>   S(cale) X(-scale) M(anual) L(ower): **-3** U(pper): **3** Q(uit)
>   F(ormat) A L(ines) Q(uit)
>   D(ata-Labels) A R(ange): **A6..A10** C(enter) Q(uit) Q(uit)

The Graph View and **<F10>** commands operate as they do in *Quattro Pro*. In Version 3.0 the graph font and format settings are adjusted through the Graph Options Advanced menu selection, and a graph is named by the Graph Name Create command. In Versions 2.2 and 2.3 many settings can be selected by entering an edit menu from within the graph menu by pressing the **<F2>** key. In the edit menu a user can select the type, range, color, orientation, grid line, or zero line (axes) options. Also, the Graph Options menu allows for the choice of legends, formats, titles, and data labels (i.e., interior labels).

As with other spreadsheets, Lotus 1-2-3 can be set to provide scales for the x- and y-axes automatically. Frequently the automatic settings will prove to be reasonably satisfactory. In addition, as in *Quattro Pro*, the scales for the x- and y-series can be set manually, as indicated above, to provide for a $x{:}y$ aspect ratio that is approximately correct. However, one difficulty with *Lotus* graphs is that the size of the increments used with the scales of the x- and y-axes cannot be set manually. As a consequence, the resulting $x{:}y$ aspect ratio may not be ideal. One way to adjust the ratio more finely is through the use of the Graph Options Titles command that creates titles, or headings, at the top of a graph. Using these headings reduces the height of the resulting graph. The headings can contain actual text, or they can simply consist of spaces that are not shown, but that take up space. By adjusting the size of the font used for the titles (with versions in which this capability is provided), the aspect ratio can be refined to attain good proportions. Other drawbacks of *Lotus* graphic capabilities lie in the inability of a user to change colors or to hide the x- and y-axes in some versions. However, some add-on packages can provide these capabilities.

With Version 3.0 a graph can be inserted into a window on the right side of the active spreadsheet, allowing both graph and numerical output to be displayed simultaneously, much as in *Quattro Pro* and *Excel*. This capability makes an effective implementation of the animation models of Chapter 4 possible, and allows all models to be made more visually oriented. In Version 3.0 a window for a graph is generated to the right side of the cursor location in the spreadsheet via the following command:

> / W(orksheet) W(indow) G(raph)

Because the graph is displayed in the narrower region provided by a partial vertical screen, it may be necessary to adjust the *x*- or *y*-scales in order to obtain an appropriate aspect ratio for the split-screen display.

In Version 2.3 a graph is inserted by first activating the WYSIWYG add-in feature through the commands: /Add-in, Attach, WYSIWYG.ADN. Next, the WYSIWYG menu is displayed by pressing the : key. The Graph Add command is selected from this menu. This command enables the graph to be inserted into the spreadsheet display with a size and a location selected by a user.

1.8. MODIFICATIONS FOR *MICROSOFT EXCEL*

Microsoft Excel is another leading spreadsheet that possesses many excellent features of its own. Although the fundamental approach to the operation of *Excel* is the same as that of other spreadsheets, some specific implementations and commands are quite distinctive. This section discusses some of the primary differences that arise in creating graphs of curves in *Excel*. Books and manuals should be consulted for other specific details.

1.8.A. Spreadsheet Conventions

In *Excel*, formulas, functions, and expressions that produce values are indicated by an initial = sign, and not by using such initial symbols as +, number, (, or @. Thus, the expression =B$4*B7 is used rather than +B$4*B7, and =COS(D4) is employed instead of @COS(D4). Conventions for the functions that use column or row counts differ slightly as well. Whereas *Quattro Pro* starts the count of columns in a table with the left-most column as column 0, *Excel* starts this same count at 1. Thus, the column references listed in the book for VLOOKUP and CHOOSE functions must be increased by 1 for *Excel*. Also, *Excel* provides for table names as absolute locations. Thus, instead of *Quattro Pro's* @VLOOKUP(B4,$T,5), where T is the range B12..G85, *Excel* uses =VLOOKUP(B4,T,6) or =VLOOKUP(B4,$B$12:$G$84,6). Notice that ranges are denoted by B12:G65.

Excel operations and commands are primarily designed for the use of a mouse. As one example, the mouse can be used to find and select functions from a menu by using the Formulas Paste Functions option.

Another difference in using *Excel* occurs in the copy procedure. Although *Excel* has a copy command, the copy technique required in this book's examples is implemented best in *Excel* via the Fill command. For example, suppose that it is desired to copy the formulas in Cells B13..D13 down their columns through Row 84. The cursor is first moved to Cell B13. Then the range B13..D84 is selected by holding down the mouse button and dragging the cursor. Finally, the Formula Fill Down option is chosen from the menu to complete the copying procedure. The standard conventions regarding $ references are employed in *Excel*.

1.8.B. Graph Conventions

Graph option selections in *Excel* are also oriented to the use of a mouse, although frequently keyboard commands can be used as well. Manuals should be consulted for details. In *Excel* graphs are referred to as charts.

There are two basic approaches to creating an XY graph. In the first, the graph is created in a separate file. With both the model and the graph files active, the numerical model and its graph can be displayed either individually and accessed as needed, or resized and shown simultaneously. In the second approach, the graph is embedded in the active spreadsheet model, much as in *Quattro Pro*, and the mouse used to select, resize, and move the screen display. In either approach, graph features can be adjusted by first selecting the graph and then using mouse and/or keyboard commands. Creating an imbedded graph is facilitated by selecting a special graphic toolbar consisting of option buttons to be displayed on the screen.

Now, assume that a model has been created as in Figure 1.9, with its (x,y) graph coordinates in Columns E,F of Rows 12..84. The mouse can be used to select the block E12..F84. Then the XY graph option is selected from the toolbar, with the initial column providing the x-values. Next, the mouse is employed to drag and select an area of the spreadsheet display in which the graph is to appear and pressing **<Enter>**. The graph's size and location can be readjusted later. The graph will now appear in the chosen box.

Next, the graph is selected by double clicking on the graph box. To select the color, style, and format (lines or points) for the curve, the cursor is moved to a point on the curve and the mouse is double clicked. Choices are made from the options displayed, with **<Enter>** pressed to end that step. To adjust the x-axis, the cursor is moved to a point on the x-axis, and the mouse is double clicked. Again, selections are chosen from the options provided, including the manual scale option and the appropriate settings for maximum and minimum values of x. A similar procedure is used with the y-axis. In *Excel*, the axes appear in their mathematical location rather than along edges as in *Quattro Pro*.

Once a graph has been created, a user can move back to the worksheet file, and resize the graph's box in order to attain a better aspect ratio. Also, the box can be moved to another location or be resized. It is also possible to hide the axes and produce graphs that consist of the curve. This is done by the Axes command in the Chart menu.

One important difference encountered in using *Excel* involves its "pen-up" capability. Models in this book generate @NA in order to cause gaps in *Quattro Pro* and *Lotus* graphs, to "lift the pen". Unfortunately, *Excel* simply ignores such an @NA, so the gaps are not produced, and unwanted line segments will connect the separate curves. *Excel* instead requires blank cells in order to produce such gaps. In fact, it requires that the cell be completely empty. Thus, it is not possible to use an =IF statement that returns an empty string.

One fairly easy solution is first to create a model just as described in the book. Following this, the cells that contain #N/A are erased or cleared. Although this can be tedious to do by hand, it can be carried out effectively through the use of a macro program that employs the spreadsheet's Find command. Although macros are not often used in this book, this seems to be a good place to employ one. In *Excel*, macros are created in separate macro files. Details of creating macros can be found in manuals. The macro on the next page carries out the process of erasing all cells that display NA. One way to create a macro is to put it in a macro file and name it as ^E. When the file is opened, pressing **<CTRL>-E** runs the macro.

```
=SELECT.SPECIAL(3,16,2)
=CLEAR(3)
=RETURN
```

Chapter 2

CLASSICAL CURVES

The first sections of this chapter develop a fundamental curve sketching model that is used throughout the book. This model can be used directly to generate an extensive variety of interesting curves. In addition, many classical curves are created through straightforward modifications of this basic model. These models are designed to be quite general through the effective incorporation of parameters. By changing a model's parameters, users can create an extensive collection of new and attractive curves as well.

The subsequent sections describe the construction of individual classical curves. Most sections also contain presentations of one or more additional spreadsheet models that provide a geometric construction of the curve. These models create interesting and effective graphic displays, often employing the multiple-curve technique of Section 1.6 to produce a curve as an envelope of families of lines or circles. Finally, Section 2.15 presents a creative way to use a spreadsheet's solver command to produce curves of more complex equations. As in all of the book's models, these examples are designed to allow users to vary parameters, functions, and other features to produce an extensive variety of other curves that can be investigated.

In addition to the curves provided here, a variety of books provide an abundance of other curves and descriptions of constructions to create, especially Lockwood (1961), Von Seggern (1993), Lawrence (1972), Yates (1974), and Teixeira (1971). Descriptions of the creation of curves via *Mathematica* are provided in Gray (1993). Other brief descriptions of constructions via envelopes and curve stitching are contained in Hilton (1932), Pedoe (1976), Cundy and Rollett (1961), and most calculus books. Millington (1989) provides a wealth of curve stitching examples and ideas, while Seymour (1992) furnishes a nice collection of graphic output.

2.1. PARAMETRIC EQUATIONS

2.1.A. Initial Model

Frequently a curve is defined by describing its x- and y-coordinates as functions of a parameter t, $x = f(t)$, $y = g(t)$. Thus, t might represent time and $(x,y) = (f(t),g(t))$ the position of a moving point P at time t. (See Hill, 1990, pp. 115 to 140.) The graph of Figure 2.1, created by the model of Figure 2.2, shows the curve defined by the parametric equations

$$x = (t+0.43)(t-0.8)(t-1.6)(t-3.0),\ y = (t+1.0)(t-0.6)(t-1.0)(t-3.0),\ \text{for } -0.5 \leq t \leq 3.1$$

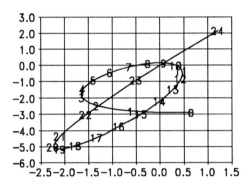

Figure 2.1. Parametric Curve.

In Figure 2.2, Column C generates t values that range from -0.50 to 3.10 in steps of size $h = 0.05$. To create the model, first the initial value of t, -0.5, is entered in Cell C12. Succeeding values of t are formed in Column C by adding h to each previous value. Thus, the formula 0.05+C12 is entered in Cell C13 and copied down Column C through Row 84. The indicated formulas for x and y as functions of t are entered in Cells D12..E12 and copied similarly to generate the (x,y)-coordinates in Columns D..E.

```
         B         C         D          E       F
10      PARAMETRIC EQUATION
11       n         t         x         y1     label
12       0      -0.50      0.669     -2.888     0
13       1      -0.45      0.177     -2.889
14       2      -0.40     -0.245     -2.856
15       3      -0.35     -0.601     -2.793     1
16       4      -0.30     -0.897     -2.703
17       5      -0.25     -1.136     -2.590
18       6      -0.20     -1.325     -2.458     2
          :
80      68       2.90     -0.909     -1.704
81      69       2.95     -0.491     -0.905    23
82      70       3.00     -0.000     -0.000
83      71       3.05      0.568      1.017
84      72       3.10      1.218      2.152    24

B12: 0         B13: 1+B12
C12: -0.5      C13: 0.05+C12
D12: (C12+0.43)*(C12-0.8)*(C12-1.6)*(C12-3)
E12: (C12+1)*(C12-0.6)*(C12-1)*(C12-3)
F12: @IF(@MOD(B12,3)=0,B12/3,"")
```

Figure 2.2. Parametric Curve: I.

Copy:		Graph:			
From	To	Series	Cells	Labels	Purpose
B13..C13	B14..C84	X	D12..D84		x-axis
D12..F12	D13..F84	1	E12..E84	F12..84	curve

Column B numbers the points of the curve. The graph of Figure 2.1 is created by designating Column D as the *x*-series and Column E as the first *y*-series. This model also uses interior labels to number every third point of the graph to illustrate the order in which the points are plotted. The label values are generated in Column F by the indicated logical @IF function. The function @IF(@MOD(B12,3)=0,B12/3,"") computes and displays the quotient $n/3$ when the value of n in Column B is a multiple of 3, and returns a blank otherwise. The graph can be enhanced further by showing the equations within the display via interior labels, as discussed in Section 1.3.

2.1.B. Extended Model

The model is made more general by incorporating parameters, as shown in Figure 2.3. While this model can be created anew, it is easier to simply make changes in the model of Figure 2.2. This practice of modifying a previously constructed spreadsheet model can be used advantageously throughout this book. In doing this, it may prove to be more efficient to set the recalculation mode to manual while the modifications are being made.

The value of h, the incremental step size for t, is set in Cell D3, while the label frequency k is entered in Cell D4. Here, as in the earlier model, $h = 0.05$ and $k = 3$. Rows 7..8 contain the parameters a_i and b_i for the formulas for x and y,

$$x = (t-a_1)(t-a_2)(t-a_3)(t-a_4), \quad y = (t-b_1)(t-b_2)(t-b_3)(t-b_4)$$

To generate the entries in Column C, the formula for $h+t$, +D$3+C12, is entered in Cell C13 and copied down the column. As in Figure 2.2, the indicated formulas for t, x, and y are entered in Cells D12..F12 and copied through Row 84.

```
        B      C      D      E      F
  1 PARAMETRIC EQUATION
  2 Overall Parameters:
  3      Step:h=   0.05
  4      Freq:k=      3
  5 Curve Parameters:
  6             P1     P2     P3     P4
  7    x:    -0.43    0.8    1.6    3.0
  8    y:    -1.00    0.6    1.0    3.0
  :
 11     n      t      x      y1  label
 12     0   -0.50  0.669  -2.9      0
 13     1   -0.45  0.177  -2.9

 B12: 0   B13: 1+B12   C12: -0.5   C13: +D$3+C12
 D12: (C12-C$7)*(C12-D$7)*(C12-E$7)*(C12-F$7)
 E12: (C12-C$8)*(C12-D$8)*(C12-E$8)*(C12-F$8)
 F12: @IF(@MOD(B12,D$4)=0,B12/D$4,"")
```

Figure 2.3. Parametric Curve: II.

Activities

1. Vary the values of the a_i, b_i in the model of Figure 2.3 to produce different curves. For example, set: (a) $b_1 = 0.3$, $b_2 = 0.2$; (b) $b_1 = -0.1$, $b_2 = -0.2$, leaving each of the other parameters fixed. Also, incorporate t_0, the initial value of t, as a parameter.
2. Experiment with different values for the label frequency k.
3. Replace the quartic polynomials for x and y by various parametric equations involving polynomial, trigonometric, exponential, and other functions. In particular, try the following functions used in catastrophe theory (Von Seggern, 1993, p. 94):
   (a) $x = c(8at^3+24t^5)$, $y = c(-6at^2-15t^4)$ for $a = -7$, $c = 0.02$, $-1.68 < t < 1.68$ (*butterfly*);
   (b) $x = c(-2at-4t^3)$, $y = c(at^2+3t^4)$ for $a = -2$, $c = 0.5$, $-1.2 < t < 1.2$ (*swallowtail*).
4. An equation of the form $y = f(x)$ can be parametrized as $x = t$, $y = f(t)$. Use this observation together with the model of this section to graph the following curves:
   (a) $y = \sin(a\pi x)\sin^{-1}(x)$, $a = 3$, $-1.2 \leq x \leq 1.2$; (b) $y = (1-2ax^2)\exp(ax^2)$, $a = -3$.
5. Generate a graph that uses two y-series to create the following curves:
   (a) $y = \pm cx(a^2-x^2)^{1/2}$ (*eight curve*); (b) $y = \pm cx^2(a^2-x^2)^{1/2}$.

2.2. PRIMARY PARAMETRIC MODEL

In this section the previous example is augmented to provide the fundamental curve sketching model upon which most of the examples in the book are based.

Many curves are defined by giving the polar coordinates (r,θ) of a point P on the curve, where r is the distance from the origin to P, and θ is the angle (in radians) that a ray from the origin to P makes with the positive x-axis (see Chapter 1 and Figure 2.4). Although the polar graph type is not provided on current spreadsheets, graphs of these functions are still easy to implement using the Cartesian (x,y)-coordinate system. Moreover, the polar grid actually can be created on a spreadsheet, as described in Section 2.3. One of the standard parametric curves is the heart-shaped *cardioid*, $r = 1 - \cos(\theta)$, shown in Figure 2.4.

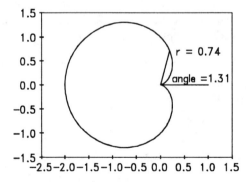

Figure 2.4. Cardioid: $r = 1 - \cos(\theta)$.

Conversion from polar to Cartesian coordinates is effected via the relations $x = r\cos(\theta)$, $y = r\sin(\theta)$, as discussed in Chapter 1. Figure 2.5 generates the graph of any curve of the more general form $r = a + b\cos(d\theta) + c\cos(e\theta)$. These formulas furnish a vast variety of interesting curves to examine. The particular example creates the cardioid of Figure 2.4.

The model generates (x,y)-coordinates for successive values of the polar angle θ. An incremental step size for θ in degrees, θ_0, is set in Cell D3. For many equations, $\theta_0 = 5°$ produces reasonably smooth curves. Values for the curve's parameters are entered in Rows 6..7. For this cardioid $a = d = 1$, $b = -1$, and $c = e = 0$. To use the trigonometric functions, each angle θ_d in degrees is changed to an angle θ_r in radians by $\theta_r = \theta_d \pi/180$. The constant $\pi/180$ is determined in Cell D9 as @PI/180. This particular calculation is used similarly throughout the book without further comment.

```
        B      C      D      E      F      G      H      I
 1 PARAMETRIC EQUATIONS:
 2 Overall Parameters:
 3  Step: θ0=    5
 4  Freq: k  =   3
 5 Curve Parameters:  r = a+bcos(dθ)+ccos(eθ)
 6        a =    1    b =   -1    c =    0
 7                    d =    1    e =    0
 8 Constants:
 9  pi/180=  0.017
10 Curve:                        Curve
11  n  θ:deg  θ:rad    r      x      y1     y2  Label
12  0    0    0.000  0.000  0.000  0.000         0
13  1    5    0.087  0.004  0.004  0.000
14  2   10    0.175  0.015  0.015  0.003
15  3   15    0.262  0.034  0.033  0.009         1
16  4   20    0.349  0.060  0.057  0.021
 :
83 71  355    6.196  0.004  0.004  0.000
84 72  360    6.283  0.000  0.000  0.000        24

D9  : @PI/180          B12: 0      B13: 1+B12
C12: +D$3*B12          D12:+D$9*C12
E12: +D$6+F$6*@COS(F$7*D12)+H$6*@COS(H$7*D12)
F12: +E12*@COS(D12)    G12: +E12*@SIN(D12)
I12: @IF(@MOD(B12,D$4),"",B12/D$4)
```

Figure 2.5. Polar Equation (Cardioid).

Copy:		Graph:			
From	To	Series	Cells	Labels	Purpose
B13	B14..B84	X	F12..F84		x-axis
C12..I12	C13..I84	1	G12..G84	I12..84	curve

The process of creating the model is straightforward. Column B counts points on the curve, $n = 0,1,2,...,72$. The other basic formulas are entered in Row 12. Column C computes the angles θ in degrees as $n\theta_0$, with Column D providing the radian equivalent. Thus, the initial angle is found in Cell C12 by +D$3*B12, and this value is converted to radians in Cell D12 by +D$9*C12. Some spreadsheets provide a library function to convert from degrees to radians, @RADIANS(C12). The corresponding value of r is computed in Cell E12 as

E12: +D$6+F$6*@COS(F$7*D12)+H$6*@COS(H$7*D12).

Next, r is used to compute the values of $x = r\cos(\theta)$ and $y = r\sin(\theta)$ in Cells F12 and G12 as +E12*@COS(D12) and +E12*@SIN(D12). Finally, the formulas of Row 12 are copied down their columns through Row 84 to compute the (x,y)-components of the points over the range $0° \leq \theta \leq 360°$.

Column I can be used to label successive points on the graph just as in the previous section, with the label frequency k set in Cell D4. Although it is not employed here, Column H is reserved to use with a second y-series to display an auxiliary curve if desired.

Once this model is created, it can be employed to generate a great variety of graphs simply by changing the values of the parameters. Representative curves are displayed in Figures 2.6 to 2.13. To modify the model for other curves of the form $r = f(\theta)$, the formula for $f(\theta)$ is entered in Cell E12 and copied down Column E. If the (x,y)-coordinates of a curve are given directly in terms of θ, the entries Column E can be erased, and the given formulas for x and y entered in Cells F12 and G12 and copied.

To improve the output it may be desirable to change the step size in Cell D3, and to insert additional rows to trace a curve fully or to produce smoother curves. This is easy to do using the standard insert and copy process discussed in Chapter 1. For example, to double the number of points for a curve, the value of Cell D3 is changed to 2.5, and the cursor moved to Row 83. Next, the Insert command is used to insert 72 additional rows. Finally, the formulas in Row 83 are copied down through Row 156. The graph is updated automatically. This process has been used in creating the graphs that are shown here and most of the others throughout the book.

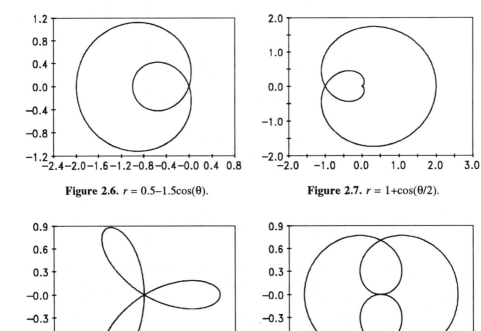

Figure 2.6. $r = 0.5 - 1.5\cos(\theta)$.

Figure 2.7. $r = 1 + \cos(\theta/2)$.

Figure 2.8. $r = \cos(3\theta)$.

Figure 2.9. $r = \cos(\theta/2)$.

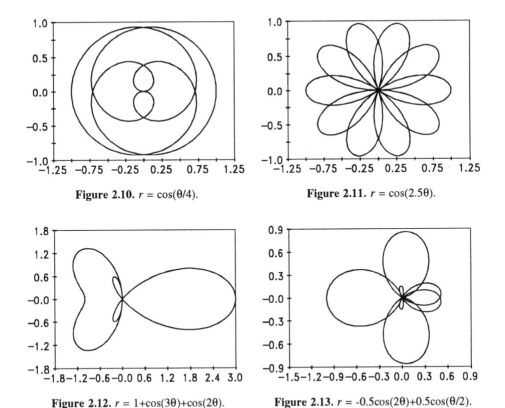

Figure 2.10. $r = \cos(\theta/4)$.

Figure 2.11. $r = \cos(2.5\theta)$.

Figure 2.12. $r = 1+\cos(3\theta)+\cos(2\theta)$.

Figure 2.13. $r = -0.5\cos(2\theta)+0.5\cos(\theta/2)$.

Activities

1. Investigate the range of curves that can be produced by varying the parameters of Figure 2.5. The following are some values to try for the parameters a through e:
   (a) 1,-1,-2,1,-3; (b) 1,-1,-2,3,3; (c) 1,-1,-2,6,3; (d) 0.5,1,1,5,3.
2. Replace the formula for r with $r = a+b\cos(d\theta)+c\sin(e\theta)$ or $r = a+b\sin(d\theta)+c\sin(e\theta)$. Compare the nature of the graphs produced.
3. Create graphs for other curves, including: (a) $r = c/\sinh(0.5\theta)$ $-4\pi < \theta < 4\pi$ (*spiral of Poinsot*); (b) $x = \cos(\theta)$, $y = \sin(\theta)\sin^n(\theta/2)$, $n = 1,2,3,...$ (*teardrop*).
4. Create the 'butterfly' curve, $r = \exp(\cos(\theta))-2\cos(4\theta)$. Then investigate the effect produced by modifying the function slightly to $r = \exp(\cos(\theta))-2\cos(4\theta)+\sin^5(\theta/12)$. Repeat using the function $r = \exp(\cos(2\theta))-1.5\cos(4\theta)$. These curves are considered in Fay (1989).
5. Investigate the rose curves, $r = \cos(n\theta)$ and $r = \sin(n\theta)$ and their generalizations. Various aspects are discussed in Althoen and Wyneken (1990) and Hall (1992).

2.3. POLAR GRAPHS

Because spreadsheets do not provide the polar graph as a standard type of display, all graphs in this book are created via the rectangular coordinate system. However, it is possible to use a spreadsheet to create a polar grid on which to display graphs of curves. The grid is formed by generating concentric circles and regularly spaced radii. Figure 2.14 shows a cardioid on a polar grid. It is generated by modifying the basic model as shown in Figure 2.15. The graphic depictions can be made even more effective through the use of colors and curve styles. It should be observed that an implementation of this model is rather large.

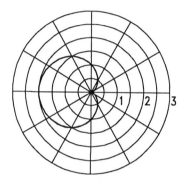

Figure 2.14. Polar Grid, $r = 1-\cos(\theta)$.

The first 84 rows of the model are identical to those of Figure 2.5. The concentric polar circles of the polar grid are created in Rows 88..530 using a technique similar to that described for generating multiple lines in Section 1.6. The construction is accomplished by entering the formulas for Rows 88..89 and copying them through Row 530 as indicated.

Column C constructs angles θ in multiples of 5° modulo 370°, with the equivalent radian measure calculated in Column D, and the x- and y-component determined in Columns F and H. When the value of a cell in Column C is 365, @NA is generated for x and y. This serves to separate the distinct circles. Column B provides the radii of the circles. The radius is initially set to 0.5 in Cell B88. Next, the formula +B88+0.5*(C89=0) is entered into Cell B89 and copied down Column B. These formulas ensure that the radius remains constant until the counter in Column C becomes 0. This signals the start of a new circle, at which time the radius is incremented by 0.5. The circles are generated by the formulas $x = r\cos(\theta)$, $y = r\sin(\theta)$, with r read from Column B and θ from Column D.

Next, the radial lines are formed in Rows 532..566 using the standard multiple line technique of Section 1.6 (except that here the roles of Columns B and C have been reversed). In Column C the angles for successive radii vary by 30°. The first three values are entered as constants in Cells C532..C534, after which a formula is used. The alternate points on each line are (0,0) and a point on the outermost circle.

Column F provides the x-axis values for the graph, Column G the y-values of the curve, and Column H the y-values of the polar grid. An effective choice of colors and curve styles can serve to highlight the curve.

```
        B      C      D     E     F     G     H
  :
  11    n    θ:deg  θ:rad   r     x    y1    y2
  12    0      0    0.00  0.00  0.00  0.00
  :
  84   72    360    6.28  0.00  0.00  0.00
  :
  87  Polar Grid (Circles):
  88  0.5     0    0.00        0.50        0.00
  89  0.5     5    0.09        0.50        0.04
  :
 160  0.5   360    6.28        0.50       -0.00
 161  0.5   365    6.37         NA          NA
 162  1.0     0    0.00        1.00        0.00
 163  1.0     5    0.09        1.00        0.09
  :
 530  3.0   360    6.28        3.00       -0.00
 531  Polar Grid (Radii):
 532    0     0    0.00        0.00        0.00
 533    1     0    0.00        3.00        0.00
 534    2    NA     NA          NA          NA
 535    0    30    0.52        0.00        0.00
  :
 564    2    NA     NA          NA          NA
 565    0   330    5.76        0.00        0.00
 566    1   330    5.76        2.60       -1.50

B88: 0.5    B89: +B88+0.5*(C89=0)
C88: 0      C89: @MOD(5+C88,370)
D88: +D$9*C88
F88: @IF(C88>360,@NA,B88*@COS(D88))
H88: @IF(C88>360,@NA,B88*@SIN(D88))
B532: 0     B533: @MOD(1+B532,3)
C532: 0 C533: 0  C534: @NA  C535: 30+C532
D532: +D$9*C532
F532: @IF(B532=0,0,B$530*@COS(D532))
H532: @IF(B532=0,0,B$530*@SIN(D532))
```

Figure 2.15. Polar Grid.

Copy:		Graph:			
From	To	Series	Cells	Labels	Purpose
B89..C89	B90..C530	X	F12..F566		x-axis
D88..H88	D89..H530	1	G12..G84		curve
B533	B534..B566	2	H12..H566		polar grid
C535	C536..C566				
D532..H532	D533..H566				

Activities

1. Implement the various curves in this chapter via polar graphs. Choose effective line styles and colors.
2. Create a spreadsheet model that displays other families of mutually orthogonal curves. For example, use $y = c_1/x$ and $y^2 - x^2 = c_2$.

2.4. CARDIOID

The *cardioid* is a heart-shaped curve defined by $r = 2a(1-\cos(\theta))$. Its graph is shown in Figure 2.4. Cardioids with other orientations are $r = 2a(1+\cos(\theta))$ and $r = 2a(1\pm\sin(\theta))$.

2.4.A. The Cardioid as an Envelope

There are many ways to construct cardioids geometrically. In this example a cardioid is formed as the envelope of a family of circles (Lockwood, 1961, p. 34). The construction starts with a given circle C and a fixed point P on C. Next, a family of circles that have centers (x_n, y_n) on C and pass through the point P is generated. Figure 2.16 is created by Figure 2.18, using the circle of radius $a = 1$ and center (0,0) as C, and $P = (x_0, y_0) = (1,0)$.

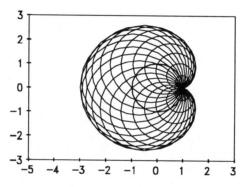

Figure 2.16. Cardioid as Envelope.

The model is made even more general by implementing the base curve C as an ellipse, $x = a\cos(\theta)$, $y = b\sin(\theta)$. The values for a and b are entered in Cells D6 and F6, and the coordinates of the fixed point $P = (x_0, y_0)$ are set in Cells G3..G4. When $a = b$, C is a circle. The figure is produced using steps of $\theta_0 = 15°$ (Cell D3) for θ. The model's curves can be made smoother by increasing the size of the model and using a smaller step size for θ.

The curve C is created in Columns B..G of Rows 12..36, just as in the basic model, with Column B providing a point count n. The radius of the first circle, whose center is at the point (x_1, y_1) that is given by Cells F12,G12, is found in Cell H12 as

$$r_1 = \sqrt{(x_1-x_0)^2+(y_1-y_0)^2}$$

This formula is copied down Column H, with the coordinates of (x_1, y_1) as relative references, and those of (x_0, y_0) as absolute. The block B12..H36 is named T. Columns F, G, and H provide the coordinates of the centers (x_n, y_n) and the radii r_n for successive circles.

These circles are generated in Rows 41..663 by using the technique of Section 2.3. The n-th circle is formed by the (x,y)-coordinates given by

$$x = x_n + r_n\cos(\alpha),\ y = y_n + r_n\sin(\alpha),\ 0 \le \alpha \le 2\pi$$

where (x_n, y_n) and r_n are the n-th centers and radii, as shown in Figure 2.17.

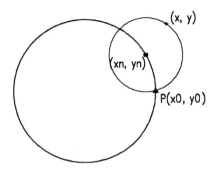

Figure 2.17. Envelope Construction.

```
         B       C       D       E       F       G       H       I       J
 1       ENVELOPE via CIRCLES (Base: Ellipse)
 2   Parameters:            Fixed Point:
 3       Step:θ0=    15      x0 =    1
 4                            y0 =    0
 5   Curve Parameters:  x = a·cos(θ), y = b·sin(θ)
 6           a =     1  b =     1
 7
 8   Constants:
 9       pi/180=0.017
10   Curve:                          Curve   Dist    Aux     Pt
11       n   θ(deg)  θ(rad)   r       x      y1      rn      y2      y4
12       0    0       0.00   1.00    0.00    0.00
13       1   15       0.26   0.97    0.26    0.26
14       2   30       0.52   0.87    0.50    0.52
  :
36      24  360       6.28   1.00   -0.00    0.00
37   Additional:
38       Fixed Point:           1                           0
39
40   Circles:        α:rad
41       1    0       0.0    1.23                           0.26
42       1    1       0.3    1.22                           0.33
43       1    2       0.5    1.19                           0.39
  :
65       1   24       6.3    1.23                           0.26
66       1   25       NA     NA                             NA
67       2    0       0.0    1.38                           0.50
68       2    1       0.3    1.37                           0.63
  :
91       2   24       6.3    1.38                           0.50
92       2   25       NA     NA                             NA
93       3    0       0.0    1.47                           0.71
  :
663     24   24       6.3    1.00                          -0.00

B12: 0    B13: 1+B12        C12: 0   C13: +C12+D$3
D12: +D$9*C12
F12: +D$6*@COS(D12)         G12: +F$6*@SIN(D12)
H12: @SQRT((F12-G$3)^2+(G12-G$4)^2)
F38: +G3   J38: +G4         B41: 1   B42: +B41+(C42=0)
C41: 0   C42: @MOD(1+C41,26)
D41: @IF(C41=25,@NA,15*D$9*C41)
F41: @VLOOKUP(B41,$T,4)+@VLOOKUP(B41,$T,6)*@COS(D41)
I41: @VLOOKUP(B41,$T,5)+@VLOOKUP(B41,$T,6)*@SIN(D41)
```

Figure 2.18. Cardioid as Envelope.

40 Cardioid

Copy:		Graph:			
From	To	Series	Cells	Labels	Purpose
B13..C13	B14..C36	X	F12..F663		x-axis
D12..H12	D13..H36	1	G12..G36		base curve
B42..C42	B43..C663	2	I12..I663		circles
D41..I41	D42..I663	3	J12..J38		fixed point

Column B counts the circles, while Column C counts the points around each circle. Column D uses Column C to compute the position angle α in radians. The formulas produce @NA for each 25th position to separate successive circles. The lookup functions in Cell F41 search the table T for values of the circle whose number is in Cell B41. The formula @VLOOKUP(B41,$T,4) finds the x-coordinate, x_n, of the circle's center, while @VLOOKUP(B41,$T,6) locates its radius, r_n. The y-coordinates are found similarly in Column I. To complete the model, the entries in Row 41 are copied through Row 663.

Finally, the graph is constructed with Column F providing the x-axis series, Column G the base curve as the first y-series, and Column I the second y-series for the family of circles.

2.4.B. Multiple Curves

Figure 2.19 shows the graphs of several cardioids $r = a(1+\cos(\theta))$ for various values of a in order to illustrate the effects of progressive changes in the parameter a. It is created by Figure 2.20, which emulates the technique used in the model above. This model's approach can be used similarly with other curves in this book.

Column B generates successive values of a, using an increment a_0 set in Cell D4. Column C counts points on each cardioid. The number of points, k, plotted for each cardioid is provided by setting the step size for θ_0 in Cell D3. For example, if $\theta_0 = 5°$, the model plots $360/5 + 1 = 73$ points numbered $0,1,\ldots,72$ and another point used for the @NA to separate successive curves. Thus, Cell F3 computes k as $2+360/D3$, giving $k = 74$.

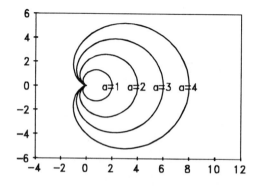

Figure 2.19. $r = a(1+\cos(\theta))$.

To create the multiple curves in Rows 8..303, +D4 and 0 are entered in Cells B8..C8 as initial values for a and n, respectively. Then the basic counting formulas are entered in

Cells B9..C9 and copied. Column D computes θ in multiples of $\theta_0\pi/180$ radians. The @IF function in Column D provides @NA appropriately. Columns E..G calculate values for r, x, and y, respectively. Column H can be used to provide interior labels for the graph.

```
        B       C    D       E      F      G   H
 1  CARDIOIDS   r = a[1+cosθ]
 2  Overall Parameters:
 3  Step:θ0=    5 Pts:k=     74
 4  Mult:a0=    1
 5  Constants:  θ0pi/180= 0.087
 6
 7   a      n   θ(rad)   r      x      y1  Lab
 8   1      0   0.00    2.00   2.00   0.00 a=1
 9   1      1   0.09    2.00   1.99   0.17
10   1      2   0.17    1.98   1.95   0.34
11   1      3   0.26    1.97   1.90   0.51
     :
80   1     72   6.28    2.00   2.00  -0.00
81   1     73    NA      NA     NA     NA
82   2      0   0.00    4.00   4.00   0.00 a=2
83   2      1   0.09    3.99   3.98   0.35
     :
154  2     72   6.28    4.00   4.00  -0.00
155  2     73    NA      NA     NA     NA
     :
230  4      0   0.00    8.00   8.00   0.00 a=4
231  4      1   0.09    7.98   7.95   0.70
     :
302  4     72   6.28    8.00   8.00  -0.00
303  4     73    NA      NA     NA     NA

F3: 2+360/D3         F5: +D3*@PI/180
B8: +D4              B9: +B8+D$4*(C9=0)
C8: 0                C9: @MOD(1+C8,F$3)
D8: @IF(C8=F$3-1,@NA,F$5*C8)
E8: +B8*(1+@COS(D8))
F8: +E8*@COS(D8)  G8: +E8*@SIN(D8)
H8: @IF(C8=0,+"a="&@STRING(B8,0),"")
```

Figure 2.20. Multiple Cardioids.

Copy:		Graph:			
From	To	Series	Cells	Labels	Purpose
B9..C9	B10..C303	X	F8..F303		x-axis
D8..H8	D9..H303	1	G8..G303	H8..H303	cardioids

<u>Activities</u>

1. Create cardioids through alternate formulas, such as $r = a(1+\sin(\theta))$. Also, modify the basic model to include a parameter α that allows a user to rotate a curve through an angle α. This can be done by replacing θ by $\theta+\alpha$ in the model's formulas.
2. Vary the values of a and b in Figure 2.18 to produce an ellipse as the base curve. Experiment with other base curves to see the resulting envelopes.

3. For any angle β in degrees, the tangents to the cardioid at the angles β, $\beta+120$, $\beta+240$ are parallel. Create a model to illustrate this fact, as shown in Figure 2.21.

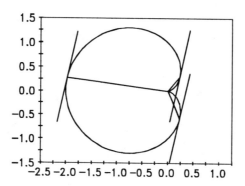

Figure 2.21. Tangents: $r = 1 - \cos(\theta)$.

4. Using the model of Figure 2.18, select the fixed point at other points on the base curve, as well as inside and outside it. Find interesting designs that are produced (Figure 2.22).
5. Modify the model of Figure 2.20 to generate multiple copies of other classes of curves, such as $r = 1 + a\cos(\theta)$, or $r = a + \cos(\theta)$, as shown for $a = 0.0, 0.4, 0.8,\ldots$ in Figure 2.23. Examine the family of teardrops of Exercise 3, Section 2.2.

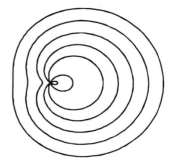

Figure 2.22. Envelope from Ellipse.   **Figure 2.23.** $r = a + \cos(\theta)$.

2.5. LIMAÇON

The *limaçon of Pascal* of Figure 2.24 can be produced from the basic parametric model. The name limaçon means "snail". A polar equation for a limaçon is $r = k + 2a\cos(\theta)$. When $k = 2a$, this determines a cardioid. The limaçon can also be formed as the envelope of circles (Figure 2.25) using the model of Figure 2.18 and choosing a fixed point external to the base circle, here P = (2,0). (See Lockwood, 1961, p. 49.) The limaçon of Figure 2.25 is a translation of that in Figure 2.24.

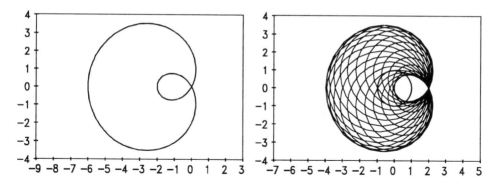

Figure 2.24. Limaçon: $r = 2-4\cos(\theta)$. **Figure 2.25.** Limaçon as Envelope.

2.5.A. Dürer's Method

The artist Dürer developed another method for drawing the limaçon. (See Lockwood, 1961, p. 49; Pedoe, 1976, pp. 230 to 231). To implement his method, points are plotted at regularly spaced intervals of $\theta_0 = 5°$ on the circumference of a circle. Starting at $\theta = 0°$ and proceeding counterclockwise, these points are numbered consecutively as $0,1,2,\ldots$. At each point n, $0 \le n \le 72$, a line segment of a fixed length k is drawn parallel to the radius drawn to point $2n$. The curve that is formed by connecting the endpoints of the successive line segments is a limaçon. Figures 2.26 and 2.27 contain the graphical output of a model for this construction, given in Figure 2.28.

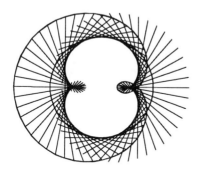
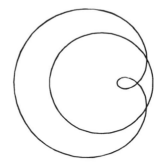

Figure 2.26. Dürer's Limaçon. **Figure 2.27.** Dürer's Limaçon without Lines.

The length, k, of the line segments and the circle radius a are set in Cells D4 and D6. The value of the position angle increment θ_0 in degrees is entered in Cell D3, while its radian equivalent is determined in Cell F9 by +D3*D9. The base circle is constructed in Cells B12..G84 using the standard procedure for creating parametric curves. The n-th point coordinates are given by $(a\cos(n\theta_0), a\sin(n\theta_0))$. Columns F and H of Rows 86..303 use these values to construct the appropriate lines using the standard multiple line technique. The alternate points are k units in either direction from the n-th point in the direction of $2n\theta_0$:

$$x = a\cos(n\theta_0) \pm k\cos(2n\theta_0), \quad y = a\sin(n\theta_0) \pm k\sin(2n\theta_0)$$

44 Limaçon

After the block B86..I303 is named S, Rows 305..377 generate points of the limaçon by reading from S the coordinates of the first endpoint of each parallel line. The indicated formulas are entered in Row 305 and copied down their columns. Another limaçon is defined by the alternate endpoints.

```
         B       C         D      E     F        G        H        I
    2 Overall Parameters:         DURER: LIMACON
    3   Step:θ0=          5
    4       k =         0.7
    5 Curve Parameters: x = a·cos(θ),  y = a·sin(θ)
    6       a =           1
    7
    8 Constants:
    9   pi/180=        0.017    s= 0.087
   10 Curve:                                 Circle    Para  Limacon
   11   n     θ:deg     θ:rad     r     x      y1        y2      y3
   12   0         0      0.00          1.00   0.00
   13   1         5      0.09          1.00   0.09
    :
   84  72       360      6.28          1.00  -0.00
   85 Parallels:
   86   0         0                    0.30           0.00
   87   0         1                    1.70           0.00
   88   0         2                    NA             NA
   89   1         0                    0.31          -0.03
   90   1         1                    1.69           0.21
   91   1         2                    NA             NA
    :
  303  72         1                    1.70          -0.00
  304 Limacon:
  305   0 .                             0.30                   0.00
  306   1                               0.31                  -0.03
    :
  377  72                               0.30                   0.00

  D9:  @PI/180      F9:  +D3*D9      B12: 0         B13: 1+B12    C12: +D$3*B12
  D12: +D$9*C12    F12: +D$6*@COS(D12)   G12: +D$6*@SIN(D12)
  B86: 0           B87: +B86+(C87=0)     C86: 0   C87: @MOD(1+C86,3)
  F86: @CHOOSE(C86,D$6*@COS(F$9*B86)-D$4*@COS(2*F$9*B86),
               D$6*@COS(F$9*B86)+D$4*@COS(2*F$9*B86),@NA)
  H86: @CHOOSE(C86,D$6*@SIN(F$9*B86)-D$4*@SIN(2*F$9*B86),
               D$6*@SIN(F$9*B86)+D$4*@SIN(2*F$9*B86),@NA)
  B305: +B12   F305: @VLOOKUP(B305,$S,4)   I305: @VLOOKUP(B305,$S,6)
```

Figure 2.28. Dürer's Limaçon.

Copy:		Graph:			
From	To	Series	Cells	Labels	Purpose
B13	B14..B84	X	F12..F377		x-axis
C12..G12	C13..G84	1	G12..G84		circle
B87..C87	B88..C303	2	H12..H303		parallels
F86..H86	F87..H303	3	I12..I377		limaçon
B305..I305	B306..I377				

2.5.B. Limaçon via Lines

There is yet another technique for generating a limaçon. (See Lockwood, 1961, p. 45.) A fixed point A is selected on the circumference of a circle C. For each point Q on C, a segment of the line determined by A and Q is created, extending k units in each direction from Q. The curve determined by the ends of these line segments is a limaçon, as shown in Figure 2.29. This construction is left as an exercise.

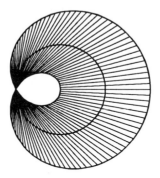

Figure 2.29. Limaçon.

Activities

1. Create a model for the construction in Part B.
2. Examine the effect of changing the location of the fixed point P in Figure 2.18 and the fixed point A in the model of Section B.
3. Create another limaçon from the alternate set of endpoints of the line segments in Dürer's model of Figure 2.28.
4. Use the multiple-curve technique of Figure 2.20 to create several copies of limaçons by using various values of a.

2.6. ASTROID

The *astroid* is a star-shaped figure defined by the parametric equations $x = a\cos^3(\theta)$, and $y = a\sin^3(\theta)$. Figure 2.30 shows the graph produced from the model of Figure 2.31, which generates a more general version of the curve, with $x = a\cos^3(c\theta)$ and $y = b\sin^3(d\theta)$.

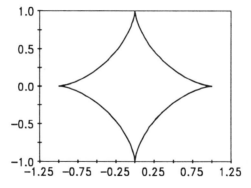

Figure 2.30. Astroid: $x = \cos^3(\theta)$, $y = \sin^3(\theta)$.

46 *Astroid*

To create the model, parameter values $a = b = c = d = 1$ are set in Rows 6..7. Columns B..D are formed as in the model of Section 2.2. Column E is left blank, and the equations for x and y are entered directly into Cells F12..G12 and copied. To trace out the entire astroid, θ extends over the interval $0° \leq \theta \leq 720°$. For an increment of $\theta_0 = 5°$ (Cell D3), using the insert and copy technique of Chapter 1 extends the formulas through Row 156.

```
            B        C       D      E     F       G
 1  ASTROID
 2  Overall Parameters:
 3     Step:θ0=     5
 4
 5  x= a[cos(cθ)]^3, y= b[sin(dθ)]^3
 6         a =      1 b =    1
 7         c =      1 d =    1
 8  Constant:
 9       pi/180=0.017
10  Curve:                               Curve
11     n   θ:deg  θ:rad   r       x       y1
12     0       0   0.00        1.00    0.00
13     1       5   0.09        0.99    0.00
14     2      10   0.17        0.96    0.01
15     3      15   0.26        0.90    0.02
     :
156  144     720  12.57        1.00   -0.00

B12: 0                      B13: 1+B12
C12: +B12*D$3               D12: +D$9*C12
F12: +D$6*@COS(D$7*D12)^3
G12: +F$6*@SIN(F$7*D12)^3
```

Figure 2.31. Astroid.

Copy:		Graph:			
From	To	Series	Cells	Labels	Purpose
B13	B14..B156	X	F12..F156		x-axis
C12..G12	G13..G156	1	G12..G156		astroid

By varying the parameter values of Figure 2.31, a diverse variety of curves can be produced, two of which are illustrated in Figures 2.32 and 2.33. In creating such curves it is often advantageous to change the step size θ_0, increase the number of points, or do both.

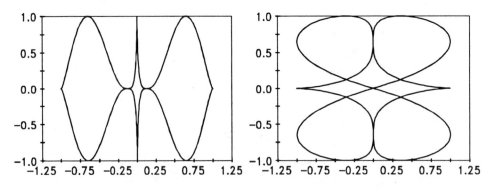

Figure 2.32. $x = \cos^3(\theta/3)$, $y = \sin^3(\theta)$.   **Figure 2.33.** $x = \cos^3(\theta/2)$, $y = \sin^3(\theta/3)$.

2.6.A. Astroid: Envelope of Lines: I

There are several ways to generate an astroid as an envelope of a family of curves. In the first example, lines are employed in a construction that occurs in the art of curve stitching, or string-and-nail design. (See Lockwood, 1961, pp. 52 to 53; Millington, 1989, p. 51.) On the x- and y-axes, pairs of points (the nails) such as (8,0) and (0,0), (7.5,0) and (0,0.5), and (7,0) and (0,1), are connected with lines (the strings). Here, all of the components of pairs of points (x,y), for $-8 \leq x \leq 8$, $-8 \leq y \leq 8$, sum to 8 and produce Figure 2.34. While it is easy to enter the coordinates of the endpoints manually, it can be challenging to do so via formulas. Although the details are not discussed here, the formulas shown in Figure 2.35 accomplish the task, producing the graph of Figure 2.34.

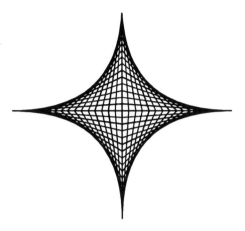

Figure 2.34. Astroid via Lines: I.

```
          B        C           D          E
   3           Step:x0=       0.5
   4             Sum=          8
  11           n       mod     x          y1
  12           0        0      NA         NA
  13           0        1      0.0        0.0
  14           0        2      8.0        0.0
  15           1        0      NA         NA
  16           1        1      0.0        0.5
  17           1        2      7.5        0.0
  18           2        0      NA         NA
  19           2        1      0.0        1.0
  20           2        2      7.0        0.0
   :
 205          64        1      0.0        0.0
 206          64        2      8.0        0.0

 B12: 0        B13: +B12+(C13=0)
 C12: 0        C13: @MOD(1+C12,3)
 D12: @NA      D13: 0            D14: +D4
 E12: @NA      E13: 0            E14: 0
 D15: +D12     D16: +D13         D17: +D4-D3
 E15: +E12     E16: +E13+D3      E17: 0
 D18: @CHOOSE(C18,@NA,0,
   @IF(@ABS(D15)=D$4,D12,2*D15-D12))
 E18: @CHOOSE(C18,@NA,
   @IF(@ABS(E15)=D$4,E12,2*E15-E12),0)
```

Figure 2.35. Astroid via Lines.

48 Astroid

Copy:		Graph:			
From	To	Series	Cells	Labels	Purpose
B13..C13	B14..C206	X	D12..D206		x-axis
D18..E18	D19..E206	1	E12..E206		lines

The x-step size x_0 is set in Cell D3 and the constant sum in Cell D4. If a value other than 8 is used for the sum, then it may be necessary to insert or delete rows. The model's layout employs the standard multiple line construction of Section 1.6. The alternate points in each pair are the x- and y-intercepts. The expressions for the first two lines are entered individually in Rows 12..17. The formulas shown for Row 18 are then entered and copied.

2.6.B. Astroid: Envelope of Ellipses

The model of Figure 2.37 forms an astroid as the envelope of the family of ellipses given by $x = a\cos(\theta)$, $y = (S-a)\sin(\theta)$, that are generated for successive values of a as a varies from a_0 (here, 0.5) to S (here, 7.5) in incremental steps of size a_0 (Cell D4). The graph produced is shown in Figure 2.36. The step size for θ, $\theta_0 = 15°$, is set in Cell D3. This model presents another variation of the basic multiple-curve technique. Here, the angle measure, rather than a point count, is used to identify points on each curve. Cell F3 creates the angle size of the "extra" angle to be used for the @NA curve separator, as $360+\theta_0$, while Cell F4 creates the modulus used to reset new curves as $360+2\theta_0$.

The multiple curve sketching scheme is employed in Rows 12..401. Column B creates the value of a for each ellipse. The value for a is initially set to 0.5 in Cell B12. This value is constant until it is incremented by the amount a_0 (Cell D4) to produce a new ellipse whenever the entry in Column C becomes 0. Column C generates the angles for points on each ellipse in increments of θ_0 (Cell D3) modulo F$4. The formulas in Column D return @NA for the points whose angle equals the value of Cell F$3, and the radian angle measurement for each point otherwise. The formulas for $x = a\cos(\theta)$ and $y = (S-a)\sin(\theta)$ are entered as +B12*@COS(D12) and (D$6-B12)*@SIN(D12) in Cells E12..F12. Then the formulas in Cells D12..F12 are copied down through Row 401.

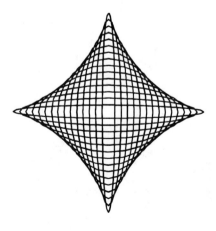

Figure 2.36. Astroid via Ellipses.

```
     B       C        D         E         F
 1 ASTROID: ENVELOPE (via ellipses)
 2 Overall Parameters:
 3     Step:θ0=       15 Limit =       375
 4     Step:a0=      0.5 Mod   =       390
 5 Curve Parameters:
 6     Sum,S =         8
 7
 8 Constants:
 9     pi/180= 0.017
10 Curve:                                Curve
11    a  θ(deg)   θ(rad)        x         y1
12  0.5        0   0.000    0.500      0.000
13  0.5       15   0.262    0.483      1.941
14  0.5       30   0.524    0.433      3.750
 :
36  0.5      360   6.283    0.500     -0.000
37  0.5      375      NA       NA         NA
38  1.0        0   0.000    1.000      0.000
39  1.0       15   0.262    0.966      1.812
40  1.0       30   0.524    0.866      3.500
 :
62  1.0      360   6.283    1.000     -0.000
63  1.0      375      NA       NA         NA
64  1.5        0   0.000    1.500      0.000
 :
400 7.5      360   6.283    7.500     -0.000
401 7.5      375      NA       NA         NA

F3:  360+D3       F4:  +F3+D3
B12: +D4          B13: +B12+D$4*(C13=0)
C12: 0            C13: @MOD(C12+D$3,F$4)
D12: @IF(C12=F$3,@NA,D$9*C12)
E12: +B12*@COS(D12)
F12: (D$6-B12)*@SIN(D12)
```

Figure 2.37. Astroid via Ellipses.

Copy:		Graph:			
From	To	Series	Cells	Labels	Purpose
B13..C13	B14..C401	X	E12..E401		x-axis
D12..F12	D13..F401	1	F12..F401		ellipses

2.6.C. Astroid: Envelope of Lines: II

There is yet another clever way of forming an astroid as the envelope of a family of lines. (See Lockwood, 1961, pp. 53 to 55.) In this scheme points are counted in two ways. First, points are numbered in a counterclockwise fashion around the circumference of a circle, starting at 0° and using increments of 5°. At the same time, for a positive integer m, another count starts at 180° in increments of $5m°$ in the clockwise direction. Line segments of length k are constructed through the n-th pairs of corresponding points. A model for this construction is presented in Figure 2.40 without extended discussion. In the figure, the pairs (x,y_1) represent the counterclockwise points, while the $(x\#,y\#)$ are the corresponding clockwise points. Columns J..K determine the coordinates of one set of line segment

50 *Astroid*

endpoints and Columns L..M the other. These are determined from the fact that the endpoints of the desired line segment determined by the points (x_0,y_0) and (x_1,y_1) are $(x_0 \pm k(x_1-x_0), y_0 \pm k(y_1-y_0))$.

When $m = 3$, an astroid is produced as in Figure 2.38, while if $m = 2$ a deltoid results (see Figure 2.46). The completion of this construction is left as an exercise. By connecting the endpoints of the line segments, other interesting curves are formed, as in Figure 2.39.

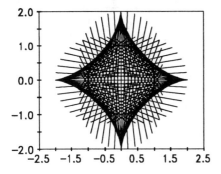

Figure 2.38. Astroid via Lines: II.   **Figure 2.39.** Astroid Embellished.

```
         B    C    D    E    F      G        H      I     J     K     L     M     N
 1 ASTROID
 2 Parameters:
 3    Step:θ0=   3           m =     3
 4                           k =    0.5
 5 Curve Parameters: x = a·cos(θ),  y = a·sin(θ)
 6       a =     1
 7
 8 Constants:
 9 pi/180= 0.017
10 Curve:                       Curve     Opposite                              Aux
11       n   θ:d  θ:rad    x      y1      x#     y#    xlo   ylo   xhi   yhi    y2
12       0    0   0.00    1.00   0.00    -1.0    0.0   2.0  -0.0  -2.0   0.0
13       1    3   0.05    1.00   0.05    -1.0    0.2   2.0   0.0  -2.0   0.2
14       2    6   0.10    0.99   0.10    -1.0    0.3   2.0   0.0  -1.9   0.4
    :
132    120  360   6.28    1.00  -0.00    -1.0   -0.0   2.0  -0.0  -2.0  -0.0
133 Lines:
134      0    0           2.00                                                -0.00
135      0    1          -2.00                                                 0.00
136      0    2            NA                                                   NA
137      1    0           1.99                                                 0.00
138      1    1          -1.98                                                 0.21
139      1    2            NA                                                   NA
    :
494    120    0           2.00                                                -0.00
495    120    1          -2.00                                                -0.00
496    120    2            NA                                                   NA

D9:   @PI/180      B12: 0     B13: 1+B12        C12: 0     C13: +C12+D$3
D12:  +D$9*C12     F12: +D$6*@COS(D12)          G12: +D$6*@SIN(D12)
H12:  +D$6*@COS(@PI-G$3*D12)                    I12: +D$6*@SIN(@PI-G$3*D12)
J12:  +F12-G$4*(H12-F12)                        K12: +G12-G$4*(I12-G12)
L12:  +H12-G$4*(F12-H12)                        M12: +I12-G$4*(G12-I12)
B134: 0    B135: +B134+(C135=0)        C134: 0    C135: @MOD(1+C134,3)
F134: @CHOOSE(C134,@VLOOKUP(B134,$T,8),@VLOOKUP(B134,$T,10),@NA)
N134: @CHOOSE(C134,@VLOOKUP(B134,$T,9),@VLOOKUP(B134,$T,11),@NA)
```

Figure 2.40. Astroid (Method C).

Copy:		Graph:			
From	To	Series	Cells	Labels	Purpose
B13..C13	B14..C132	X	F12..F496		x-axis
D12..M12	D13..M132	1	G12..G132		circle
B135..C135	B136..C496	2	N12..N496		lines
F134..N134	F135..N496				

Activities

1. Vary the parameters of the model of Figure 2.31 to generate other curves. Replace the exponents of 3 by odd integer parameters m,n to generate sketches of the parametric curves $x = a\cos^m(c\theta)$, $y = b\sin^n(d\theta)$ as in Figure 2.41.
2. Implement the model for Part C. Include the curves that are formed by the line segment endpoints. Use color effectively. Experimentally find k so that the endpoints lie on the astroid.

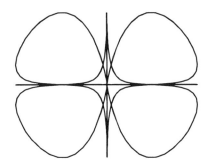
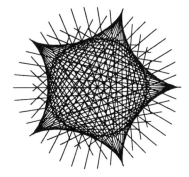

Figure 2.41. $x = \cos^9(0.6\theta)$, $y = \sin^9(0.8\theta)$.   **Figure 2.42.** Envelope of Lines: II, m = 4.

3. Vary the value of m in the model of Figure 2.40 (see Figure 2.42).
4. Design additional curve stitching models, as in Figure 2.43. Millington (1989) provides a wealth of additional ideas, including many of the curves presented in this chapter.

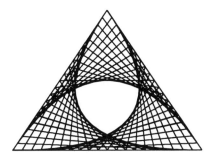

Figure 2.43. Parabolas from Triangle.

2.7. DELTOID

The *deltoid* is a Δ-shaped curve given by the parametric equations

$$x = a(2\cos(\theta)+\cos(2\theta)), \quad y = a(2\sin(\theta)-\sin(2\theta))$$

Figure 2.44 shows the deltoid with $a = 1$ and its circumscribed circle. This curve is implemented in Figure 2.45 as a modification of the basic parametric model through the more general curve $x = a(d\cos(b\theta)+e\cos(c\theta))$, $y = a(f\sin(b\theta)+g\sin(c\theta))$. The parameters shown in Rows 6..7 generate the graph of Figure 2.44. This model also can be used to generate the nephroid of Section 2.8 and other curves.

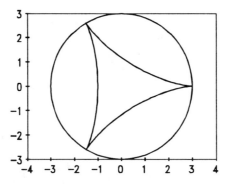

Figure 2.44. Deltoid.

```
           B       C        D     E    F       G      H        I       J
 1 PARAMETRIC EQUATIONS: Deltoid/Nephroid
 2 Overall Parameters:                              Circle:
 3    Step:θ0=    5                                    r =         3
 4
 5 x= a[d·cos(bθ)+e·cos(cθ)],  y= a[f·sin(bθ)+g·sin(cθ)]
 6        a =       1  b =    1  d =     2  f =         2
 7                     c =    2  e =     1  g =        -1
 8 Constants:
 9      pi/180= 0.017
10 Curve:                              Curve Circle
11     n  θ(deg) θ(rad)    r      x     y1    y2
12     0       0  0.000         3.00   0.00
13     1       5  0.087         2.98   0.00
14     2      10  0.175         2.91   0.01
15     3      15  0.262         2.80   0.02
 :
84    72     360  6.283         3.00   0.00
 :
88 Circle:
89     0       0   0.00         3.00          0.00
90     1       5   0.09         2.99          0.26
 :
161   72     360   6.28         3.00         -0.00

B12: 0   B13: 1+B12   C12: +D$3*B12   D12: +D$9*C12
F12: +D$6*(H$6*@COS(F$6*D12)+H$7*@COS(F$7*D12))
G12: +D$6*(J$6*@SIN(F$6*D12)+J$7*@SIN(F$7*D12))
D9:  @PI/180   B89: +B12   C89: 5*B89   D89: +D$9*C89
F89: +J$3*D$6*@COS(D89)    H89: +J$3*D$6*@SIN(D89)
```

Figure 2.45. Deltoid.

Copy:		Graph:			
From	To	Series	Cells	Labels	Purpose
B13	B14..B84	X	F12..F161		x-axis
C12..G12	C13..G84	1	G12..G84		deltoid
B89..H89	B90..H161	2	H12..H161		circle

The deltoid is created in Cells B12..G84. The formulas for x and y are entered directly in Cells F12 and G12 and copied down their columns. Rows 89..161 create a circle of radius a as an auxiliary curve in Columns F and H. The indicated formulas need only be entered in Row 89 and copied.

The deltoid can also be generated as an envelope of lines, as in Figure 2.46, by setting $m = 2$ in the model of Figure 2.40 of Section 2.6. (See Lockwood, 1961, p. 75.)

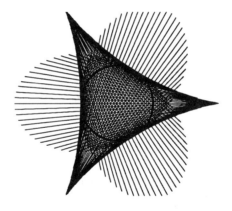

Figure 2.46. Deltoid via Lines.

Activities

1. Experiment with values for the parameters of the model of Figure 2.45 to produce other curves.
2. Using α as a parameter, modify the model and its equations in order allow for the rotation of the deltoid through an angle α. Next, select α so that one of the cusps of the deltoid appears at the top of the display, or in any other selected orientation. Also, allow the deltoid to be translated so that its center is at (a,b), where a and b are parameters.
3. Design a model that creates a deltoid via Simpson's line. (See Pedoe, 1976, p. 237; Lockwood, 1961, pp. 72 to 73.) Starting with a point P on the circle that circumscribes a triangle XYZ, draw the perpendicular lines PU, PV, PW to the respective triangle sides YZ, ZX, XY. Then draw the resulting lines UVW as P varies around the circle.

2.8. NEPHROID

The *nephroid* is a kidney-shaped curve defined by

$$x = a(3\cos(\theta) - \cos(3\theta)), \quad y = a(3\sin(\theta) - \sin(3\theta))$$

The graph of Figure 2.47 is created from the deltoid model of Figure 2.45 by setting the parameters $a = b = 1$, $e = g = -1$ and $d = f = c = 3$ in the formulas

$$x = a(d\cos(b\theta) + e\cos(c\theta)), \quad y = a(f\sin(b\theta) + g\sin(c\theta))$$

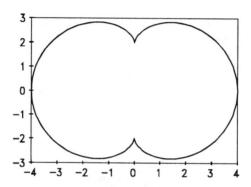

Figure 2.47. Nephroid.

2.8.A. Nephroid as an Envelope of Circles

The nephroid can be created as an envelope of a family of circles, as in Figure 2.48. The model of Figure 2.49 constructs the family of circles whose centers lie on a given base curve C, with each circle tangent to the line $x = c$. This model uses C as an ellipse, with $x = a\cos(\theta)$, $y = b\sin(\theta)$. The values for a and b are entered in Cells D6 and F6. A circle is produced by setting $a = b = 0.8$, while setting $c = 0$ establishes the y-axis as the tangent line. The envelope of these circles is the nephroid of Figure 2.48. (See Lockwood, 1961, pp. 62 to 63.)

Figure 2.48. Nephroid via Circles.

The base curve C is created in Cells B12..G36 by the usual method, but using steps of $\theta_0 = 15°$ set in Cell D3. Smoother curves can be produced by setting Cell D3 to 5°. However, additional rows must then be inserted to form all of the circles, so the size of the resulting model is tripled. The initial point (x_1, y_1) on the base curve is formed in Cells F12 and G12 using $x = a\cos(\theta)$, $y = b\sin(\theta)$. Cell H12 determines the distance d_1 from (x_1, y_1) to the line $x = c$, or $d_1 = |x_1 - c|$, by @ABS(F12-G$3). The formulas in Row 12 are copied through Row 36, with the block B12..H36 named T. For each n, the points (x_n, y_n) in Columns F and G of T (the fourth and fifth columns of T) are the centers of the n-th circle, while the distances d_n in Column H (the sixth column of T) provide their radii. Finally, Rows 40..662 generate the circles using lookup functions, just as in the cardioid example of Figure 2.18. In this block, Column B provides a count for the circles, Column C a count of points on circles, and Column D the angles α of points. The (x,y)-coordinates of the nephroid are determined in Columns F and I by $x = x_n + d_n\cos(\alpha)$, $y = y_n + d_n\sin(\alpha)$.

```
         B      C        D      E      F       G       H       I
  1  ENVELOPE via CIRCLES (Base: Circle/Ellipse)
  2  Parameters:         Line x = c:
  3    Step:θ0=    15         c =      0
  4
  5  Curve Parameters:  x = a·cos(θ),  y = b·sin(θ)
  6          a =      0.8 b =  0.8
  :
 10  Curve:                          Curve   Dist    Aux
 11    n  θ(deg) θ(rad)        x      y1      d      y2
 12    0       0   0.00     0.80    0.00   0.80
 13    1      15   0.26     0.77    0.21   0.77
 14    2      30   0.52     0.69    0.40   0.69
  :
 36   24     360   6.28     0.80   -0.00   0.80
  :
 39  Circles:         α
 40    1       0    0.0     1.55                    0.21
 41    1       1    0.3     1.52                    0.41
 42    1       2    0.5     1.44                    0.59
  :
 64    1      24    6.3     1.55                    0.21
 65    1      25     NA      NA                      NA
 66    2       0    0.0     1.39                    0.40
  :
662   24      24    6.3     1.60                   -0.00

B12: 0     B13: 1+B12            C12: +D$3*B12
D12: +D$9*C12                    F12: +D$6*@COS(D12)
G12: +F$6*@SIN(D12)              H12: @ABS(F12-G$3)
B40: 1  B41: +B40+(C41=0)    C40: 0  C41: @MOD(1+C40,26)
D40: @IF(C40=25,@NA,15*D$9*C40)              D9: @PI/180
F40: @VLOOKUP(B40,$T,4)+@VLOOKUP(B40,$T,6)*@COS(D40)
I40: @VLOOKUP(B40,$T,5)+@VLOOKUP(B40,$T,6)*@SIN(D40)
```

Figure 2.49. Nephroid via Circles.

Copy:		Graph:			
From	To	Series	Cells	Labels	Purpose
B13	B14..B36	X	F12..F662		x-axis
C12..H12	C13..H36	1	G12..G36		base circle
B41..C41	B42..C662	2	I12..I662		envelope circles
D40..I40	D41..I662				

2.8.B. Nephroid as an Envelope of Lines

A nephroid can also be formed as an envelope of lines by making a minor modification in the model employed for the astroid in Figure 2.40, with the lines determined by different points (Lockwood, 1961, p. 67). In this scheme points are counted in a counterclockwise direction around the circumference of a circle in increments of 5°, starting at 0°. Next, for a chosen positive integer m, line segments that pass through the n-th and mn-th points are formed. The only changes needed in Figure 2.40 are to enter the formulas for $x = a\cos(m\theta)$, and $y = a\sin(m\theta)$ as

H12: +D$6*@COS(G$3*D12) and I12: +D$6*@SIN(G$3*D12)

and then to copy them. The graph of Figure 2.50 is produced by setting $m = 3$.

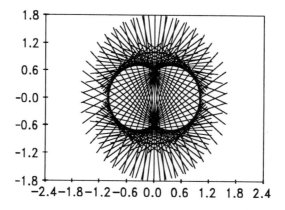

Figure 2.50. Nephroid via Lines, $m = 3$.

Figure 2.51 uses this same model with $m = 5$ to generate another curve as an envelope of lines. As in the astroid model of Figure 2.40, the modified model can be used to create new curves by connecting the ends of each of the line segments, as shown in Figure 2.52 for the curve in Figure 2.51.

Figure 2.51. Envelope, $m = 5$.

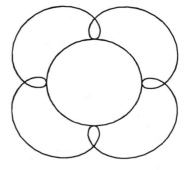

Figure 2.52. Auxiliary Curves, $m = 5$.

Activities

1. Create the model of Part B and experiment with different values of m. Modify the model to plot the graphs of curves that are formed by the endpoints of the lines segments in order to generate other interesting curves. The graph of Figure 2.53 is derived in this way from Figure 2.50.

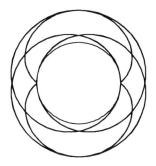
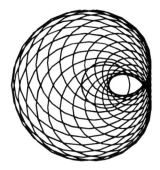

Figure 2.53. Auxiliary Curves, $m = 3$.   **Figure 2.54.** Nephroid Model, $x = 1.5$.

2. Experiment with different values for the parameters in the model of Figure 2.49 to generate some new curves. Figure 2.54 uses a base circle of radius of 0.8 with the line $x = 1.5$. Try using either a line of the form $y = mx$, or a lemniscate, as the base curve.

2.9. LEMNISCATE

The *lemniscate of Bernoulli*, with its figure eight shape, has a polar equation that is given by $r^2 = a^2\cos(2\theta)$. The graph of the equation $r^2 = \cos(2\theta)$ is shown in Figure 2.55.

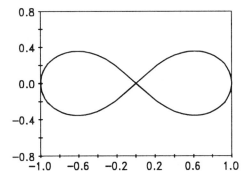

Figure 2.55. Lemniscate, $r^2 = \cos(2\theta)$.

The graph of $r^2 = a^2\cos(b\theta)$ is created by the model of Figure 2.56 as a modification of the basic parametric model. Column E computes r using

Lemniscate

$$r = |a|\sqrt{\cos(b\Theta)}$$

In this example $a = 1$ and $b = 2$. Only the positive square root is needed to generate the entire curve. To obtain a smooth graph, the step size for θ is reduced to $\theta_0 = 2.5°$ (Cell D3), and θ is extended through 720° in Rows 12..300. In the spreadsheet's numerical output, ERR is returned when the square root function encounters negative numbers. Thus, this description generates the curve in a discontinuous fashion.

```
        B         C         D         E         F         G
 1 LEMNISCATE
 2 Overall Parameters:
 3    Step: θ0=  2.5
 4
 5 Curve Parameters:   r² = a²cos(bθ)
 6        a =       1   b =       2
 7
 8 Constants:
 9    pi/180=0.017
10 Curve:                                        Lemni
11    n   θ(deg)   θ(rad)      r        x        y1
12    0    0.0     0.00     1.00     1.00     0.00
13    1    2.5     0.04     1.00     1.00     0.04
14    2    5.0     0.09     0.99     0.99     0.09
 :
30   18   45.0     0.79     0.00     0.00     0.00
31   19   47.5     0.83     ERR      ERR      ERR
 :
66   54  135.0     2.36     ERR      ERR      ERR
67   55  137.5     2.40     0.30    -0.22     0.20
68   56  140.0     2.44     0.42    -0.32     0.27
 :
300 288  720.0    12.57     1.00     1.00    -0.00

D9:  @PI/180              B12: 0       B13: 1+B12
C12: +D$3*B12             D12: +D$9*C12
E12: @ABS(D$6)*@SQRT(@COS(F$6*D12))
F12: +E12*@COS(D12)  G12: +E12*@SIN(D12)
```

Figure 2.56. Lemniscate.

Copy:		Graph:			
From	To	Series	Cells	Labels	Purpose
B13	B14..B300	X	F12..F300		x-axis
C12..G12	C13..G300	1	G12..G300		lemniscate

In using this model to sketch various lemniscates, it may be helpful to include a small positive term in the square root formula for r, e.g., @SQRT(@COS(F$6*D12)+1.0E-6). This is because the computations of the coordinates of points that should give the origin and produce θ = 0 may be rounded to return a very small negative number instead. The @SQRT would then generate ERR, thus causing an unexpected gap to appear in the graph.

2.9.A. The Lemniscate as an Envelope of Circles

Figure 2.57 shows a lemniscate as an envelope of a family of circles. This graph was created by the model of Figure 2.58. First, a hyperbola with center at (0,0) is created as the base curve. Next, the family of circles whose centers lie on the hyperbola and pass through (0,0) is formed. (See Lockwood, 1961, p. 112.)

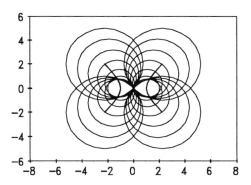

Figure 2.57. Lemniscate via Circles.

Both branches of the hyperbola, $x^2-y^2 = a^2$, are drawn in Rows 12..34. Solving for x in the equation gives

$$(1) \quad x = \pm\sqrt{a^2+y^2}$$

The value of a is set in Cell D6. Regularly spaced y-values are created in Column G using the y-step size, h, and the initial y_0 that are set in Cells D3..D4. First, y_0 is reproduced in Cell G12 by +D4. This y-value is next incremented in Cell G13 by adding h, as +G12+D$3. This formula is then copied through Row 22. Column F determines x-values on the right branch of the hyperbola using the positive square root from equation (1) in Rows 12..22. Column H calculates the distance d_n from a point (x_n,y_n) on the hyperbola to the center as

$$d_n = \sqrt{x_n^2+y_n^2}$$

The point (x_n,y_n) serves as the center of the n-th circle, with d_n as its radius. The entries of Row 12 are copied down through Row 22 to create ten points of the right branch.

The steps for creating Row 12 are repeated in Row 24, this time using the negative square root for x from equation (1). These entries are then copied through Row 34 to form another ten points on the left branch of the hyperbola. Points on the two branches are separated in Row 23 by @NA. Finally, the block B12..H34 is named T.

Rows 39..635 then create the multiple circles, as in Figure 2.18. The indicated lookup functions compute points on the n-th circle, with center at (x_n,y_n) and radius d_n, as

$$x = x_n + d_n\cos(\alpha), \ y = y_n + d_n\sin(\alpha)$$

Each circle is created from 24 points. The appearance of the resulting graph can be improved by expanding the number of points on each circle.

60 Lemniscate

```
      B   C    D   E   F      G     H    I    J
 1 LEMNISCATE (Base: Hyperbola)
 2 Parameters:
 3 Step: h= 0.4
 4 Start:y0= -2
 5 Curve Parameters: x² - y² = a²
 6    a =    1
 7
 8 Constants:
 9 pi/180= 0.02
10 Curve:              Curve Dist   Aux  Pt
11   n   θ:d  θ:r     x    y1    d   y2   y3
12   0               2.24 -2.00 3.00
13   1               1.89 -1.60 2.47
14   2               1.56 -1.20 1.97
     :
22  10               2.24  2.00 3.00
23  11                NA    NA   NA
24  12              -2.24  2.00 3.00
     :
34  22              -2.24 -2.00 3.00
35 Additional:
36    Fixed Pt:       0                 0
37
38 Circles:
39   0   0   0.0    5.24       -2.00
40   0   1   0.3    5.13       -1.22
41   0   2   0.5    4.83       -0.50
     :
63   0  24   6.3    5.24       -2.00
64   0  25    NA     NA          NA
65   1   0   0.0    4.36       -1.60
     :
635 22  24   6.3    0.76       -2.00

D9:  @PI/180                  B12: 0     B13: 1+B12
F12: @SQRT(D$6^2+G12^2)       G12: +D4   G13: +G12+D$3
H12: @SQRT(F12^2+G12^2)
F23: @NA    G23: @NA           H23: @NA
F24: -F12   G24: -G12           H24: @SQRT(F24^2+G24^2)
B39: 0   B40: +B39+(C40=0)  C39: 0  C40: @MOD(1+C39,26)
D39: @IF(C39=25,@NA,15*D$9*C39)
F39: @VLOOKUP(B39,$T,4)+@VLOOKUP(B39,$T,6)*@COS(D39)
I39: @VLOOKUP(B39,$T,5)+@VLOOKUP(B39,$T,6)*@SIN(D39)
```

Figure 2.58. Lemniscate as Envelope.

Copy:		Graph:			
From	To	Series	Cells	Labels	Purpose
B13	B14..B34	X	F12..F635		x-axis
F12,F24	F13..F22,F25..F34	1	G12..G34		hyperbola
G13,G24	G13..G22,G25..G34	2	I12..I635		envelope circles
H12	H13..H34	3	J12..J36		hyperbola center
B40..C40	B41..C635				
D39..I39	D40..I635				

Activities

1. Create the lemniscate of Figure 2.59 by using $r^2 = a^2\sin(2\theta)$.
2. Experiment with various values for a and b in Figure 2.56 (see Figure 2.60). Setting b to a fraction i/j (e.g., 2/3, 3/4, 4/5, 4,3, 3/2,...) produces a great variety of intriguing curves. However, the model will tend to miss points at the origin, resulting in gaps in the picture. One way to overcome this is to insert an extra column to the right of Column E and generate a value of r_0 to use instead of r for x and y, so that r_0 agrees with r except at the first and last cells in the sets of ERR cells. In these cases, 0 is generated. One formula that works is given by

    @IF(@ISERR(E12),@IF(@ISERR(E11)#AND#@ISERR(E13),@ERR,0),E12)

3. Modify Figure 2.58 by using a hyperbola with its center at (x_0,y_0).

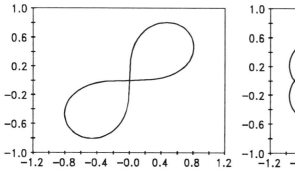
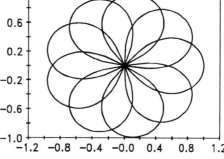

Figure 2.59. Lemniscate, $r^2 = \sin(2\theta)$.   **Figure 2.60.** $r^2 = \cos(1.8\theta)$.

2.10. CONIC SECTIONS

The conic sections can be formed by passing planes through a right circular cone. There are four basic conic types: circles, ellipses, parabolas, and hyperbolas. Equations for these curves can be presented in a number of formats (e.g., Finney and Thomas, 1990; Stewart, 1991), not all of which are easy to implement in a spreadsheet. The equations listed here can be executed readily. Models for each can be created by slight modifications in the basic model of Figure 2.5. The values of the parameters a and b are set in Cells D6 and F6. In these examples, t is used as the variable. The initial value of t is set in Cell D12. Then the designated formulas for x and y are entered in Cells F12 and G12 and copied.

Hyperbola: A hyperbola is given in Cartesian coordinates by $y^2/b^2 - x^2/a^2 = 1$. From the trigonometric identity $\tan^2(t) - \sec^2(t) = 1$, set $x = a\sec(t)$ and $y = b\tan(t)$, and enter

F12: +D$6/@COS(D12), G12: +F$6*@TAN(D12)

The two branches of the hyperbola are traced as $-\pi/2 \leq t \leq \pi/2$ (Figure 2.61).

62 Conic Sections

Ellipse: An ellipse is given by $x^2/a^2 + y^2/b^2 = 1$. From the identity $\cos^2(t) + \sin^2(t) = 1$, set $x = a\cos(t)$, and $y = b\sin(t)$, and enter

F12: +D$6*@COS(D12), G12: +F$6*@SIN(D12)

The ellipse is traced as $0 \le t \le 2\pi$ (Figure 2.62). A circle results if $a = b$.

Parabola: A parabola is given by $y = ax^2$. Because $\tan(t)$ takes on all real values in the range $-\pi/2 < t < \pi/2$, set $x = \tan(t)$, $y = a\tan^2(t)$, and enter

F12: @TAN(D12), G12: +D$6*@TAN(D12)^2

The parabola is traced as $-\pi/2 < t < \pi/2$ (Figure 2.63).

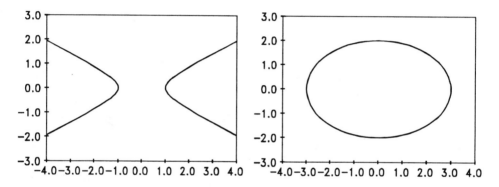

Figure 2.61. Hyperbola: $x = \sec(t)$, $y = 0.5\tan(t)$.   **Figure 2.62.** Ellipse: $x = 3\cos(t)$, $y = 2\sin(t)$.

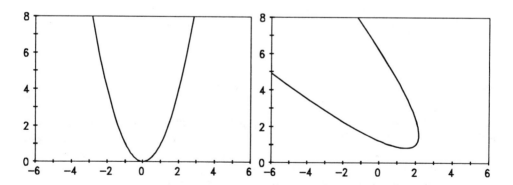

Figure 2.63. Parabola: $x = \tan(t)$, $y = \tan^2(t)$.   **Figure 2.64.** Rotation and Translation.

Activities

1. The parameter t in the equations above is not the same as the θ of polar equations. An equation for the polar form of a conic section is $r = ae/(1+e\cos(\theta))$, $e > 0$. The constant e is called the *eccentricity*. If $0 < e < 1$, the curve is an ellipse; if $e = 1$, it is a parabola; if $e > 1$, it is a hyperbola. (See Finney and Thomas, 1990, p. 696; Stewart, 1991, p. 558.) Design a model to implement this definition with e as a parameter. An alternate equation for the conic curves is $r = ae/(1-e\cos(\theta))$.

2. Implement models for each of the conic curves listed in this section, showing centers, foci, and directrices. Include interior labels to number points to show the order in which they are plotted. Experiment by changing the parameters.
3. Design a model for the polar equation of a line as $r\cos(\theta-\theta_0) = r_0$.
4. Augment the conic models to allow for the curves to be rotated through an angle α and translated by h and k units in the x- and y-directions. In Figure 2.64 the parabola of Figure 2.63 is first rotated by 45° and then translated by $(x,y) = (2,1)$ units.
5. Create conic curve stitching models. See Millington (1989) for various approaches.
6. Create a multiple curve model for a family of conics by varying a parameter.
7. From an external point P, line segments are drawn to regularly spaced points on a line L. The perpendicular bisectors of these lines form an envelope for the parabola with focus P and directrix L. Design a model to construct the lines for the envelope.

2.11. LISSAJOUS FIGURES

Lissajous, or *Bowditch*, *figures* are generated by the parametric equations $x = a\sin(b\theta)$, $y = c\sin(d\theta)$ (Taylor, 1992). These produce an extensive range of interesting closed curves, as in Figure 2.65. The graph of Figure 2.65 is produced by the model of Figure 2.66.

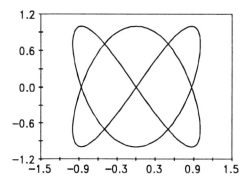

Figure 2.65. Lissajous: $x = \sin(2\theta)$, $y = \sin(3\theta)$.

```
         B       C       D       E       F       G
  2 Overall Parameters:
  3     Step:θ0=   2.5
  4 LISSAJOUS (Bowditch) Curves
  5 x = asin(bθ), y = csin(dθ)
  6         a =      1  c =      1
  7         b =      2  d =      3
  8 Constants:
  9     pi/180=0.017
 10 Curve:                               Curve
 11   n  θ(deg)  θ(rad)   r       x       y1
 12   0    0.0    0.00          0.00    0.00
 13   1    2.5    0.04          0.09    0.13
 14   2    5.0    0.09          0.17    0.26
    :
156 144  360.0    6.28         -0.00   -0.00

D9:  @PI/180       B12: 0    B13: 1+B2
C12: +D$3*B12      F12: +D$6*@SIN(D$7*D12)
D12: +D$9*C12      G12: +F$6*@SIN(F$7*D12)
```

Figure 2.66. Lissajous Figure.

64 Lissajous Figures

Copy:		Graph:			
From	To	Series	Cells	Labels	Purpose
B13	B14..B156	X	F12..F156		x-axis
C12..G12	C13..G156	1	G12..G156		curve

In Figure 2.66 the standard parametric model is employed by setting the curve parameters in Rows 6..7, and entering the equations for x and y in Cells F12 and G12 and then copying them down their columns. Because many of these curves do not repeat until θ reaches at least 720°, and as it may be desirable to assign a smaller step size θ_0 to produce smoother curves, the number of rows is extended to at least $n = 144$, as in Figure 2.66.

A wonderful variety of other curves can be produced by small changes in the model. Figures 2.67 and 2.68 are produced from the parametric equations $x = \sin^n(b\theta)$, $y = \sin^n(d\theta)$. In graphing these it is helpful to reduce θ_0 and increase the number of points plotted.

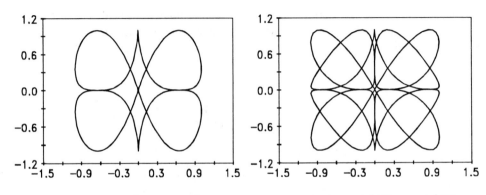

Figure 2.67. $x = \sin^3(2\theta)$, $y = \sin^3(3\theta)$.   **Figure 2.68.** $x = \sin^3(4\theta)$, $y = \sin^3(5\theta)$.

Activities

1. Create other Lissajous figures when the parameters b and d are integers, as in Figures 2.69 and 2.70. Try pairs of the form $(b, b+1)$, as well as the following pairs for (b,d): (100,150), (222,111), (70,35), (105,35), (250,500), (55,33), (33,77).

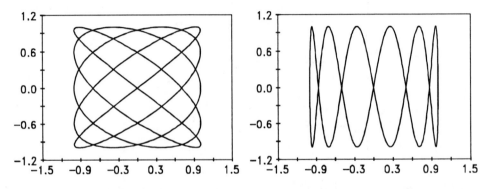

Figure 2.69. $x = \sin(5\theta)$, $y = \sin(4\theta)$.   **Figure 2.70.** $x = \sin\theta$, $y = \sin(6\theta)$.

2. Construct additional curves that are similar to the Lissajous figures. For example, set $x = \sin^n(c\theta)$, $y = \sin^n(d\theta)$, for $n = 2,3,4,5,\ldots$.
3. Create a graph for the *serpentine curves* and experiment with their parameters. Investigate the number of humps that result from various choices. These curves are given by the parametric equations $x = a\cot(ct)$, $y = b\sin(dt)$. Figure 2.71 shows the graph of $x = \cot(t/4)$, $y = \sin(4t)$.

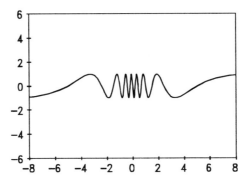

Figure 2.71. Serpentine Curve.

2.12. SPIRALS

There are several interesting spirals that can be created and investigated on a spreadsheet. The formulas for one of these models are listed here, with the others left as exercises.

2.12.A. Spiral of Archimedes

The *spiral of Archimedes* is given by $r = a\theta$. The graph of the curve $r = \theta$ is shown in Figure 2.72.

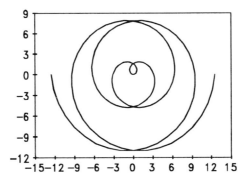

Figure 2.72. Spiral of Archimedes, $r = \theta$.

In Figure 2.73 the value of a is set in Cell D6. The graph starts at $\theta_s = -720°$ (Cell D4) and proceeds in steps of $\theta_0 = 10°$ (Cell D3) to produce Figure 2.72. To generate the θ

66   *Spirals*

values in degrees in Column C, θ_s (+D4) is reproduced in Cell C12 and $\theta + \theta_0$ (+C12+D$3) is entered in Cell C13. The latter formula is then copied through Row 156. The radian measures are determined in Column D. The formulas for $r = a\theta$ (+D$6*D12), $x = r\cos(\theta)$, and $y = r\sin(\theta)$ are next placed in Cells E12..G12 and copied down their columns. To show more of the spiral, θ can be extended in each direction.

```
         B         C      D         E        F       G
    1  ARCHIMEDES' SPIRAL
    2  Overall Parameters:
    3    Step:θ0=       10
    4    Init:θs=     -720
    5  Curve Parameters:  r = a·θ
    6      a =         1
    :
   11    n    θ:d    θ:rad       r        x       y1
   12    0   -720   -12.57   -12.57   -12.57   -0.00
   13    1   -710   -12.39   -12.39   -12.20   -2.15
    :
  156  144    720    12.57    12.57    12.57   -0.00

  D9:  @PI/180                B12: 0      B13: 1+B12
  C12: +D4                    C13: +C12+D$3
  D12: +D$9*C12               E12: +D$6*D12
  F12: +E12*@COS(D12)         G12: +E12*@SIN(D12)
```

Figure 2.73. Spiral of Archimedes.

Copy:		Graph:			
From	To	Series	Cells	Labels	Purpose
B13..C13	B14..C156	X	F12..F156		x-axis
D12..G12	D13..G156	1	G12..G156		spiral

2.12.B. Additional Spirals

There are many other spirals (Lockwood, 1961; Yates, 1974). Some of the following spirals can be created in two segments. (a) *Logarithmic (equiangular) spiral*: $r = ae^{b\theta}$ (Figure 2.74); (b) *Hyperbolic (reciprocal) spiral*: $r = a/\theta$ (Figure 2.75); (c) *Parabolic spiral*: $(r-a)^2 = b^2\theta$ (Figure 2.76 is generated by solving for r and then using both of the $\pm$ roots); (d) *Lituus*: $r^2 = a\theta$ (Figure 2.77 uses both of the $\pm$ square roots of $a\theta$ in forming r).

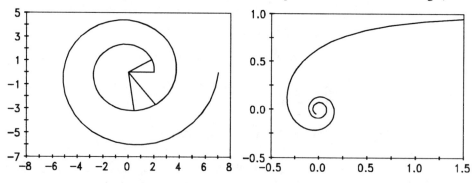

Figure 2.74. Logarithmic Spiral, $r = 2\exp(\theta/10)$.   **Figure 2.75.** Hyperbolic Spiral, $r = 1/\theta$.

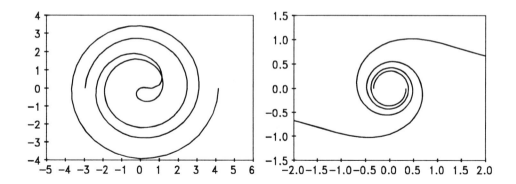

Figure 2.76. Parabolic Spiral, $(r-0.6)^2 = \theta$.   **Figure 2.77.** Lituus, $r^2 = 1.4\theta$.

Activities

1. Study the effect of setting $a < 0$ in the spiral of Archimedes.
2. Create models for each of the spirals listed above, and study the effects of changes in the parameters.
3. Implement examples of other spirals. Examples include the *epi spiral*, $r = c/\cos(n\theta)$, and *Poinsot's spirals*, $r = a/\cosh(n\theta)$ and $r = a/\sinh(n\theta)$.
4. Create the *square root spiral*. First, draw a line segment from $(0,0)$ to $(a,0)$. Then, create a right triangle by drawing a leg of length b perpendicular to the segment at $(0,a)$ and the hypothesis from the origin. Construct another right triangle using the previous hypotenuse as its base by drawing a leg of length b perpendicular to its end and its hypotenuse from the origin. Repeat the process. (See Larson et al., 1993.)

2.13. CYCLOIDS AND RELATED CURVES

A *cycloid* is the path that is traced by a point on the circumference of a circle of radius a as the circle rolls on a straight line without slipping. The curve is shown in Figure 2.78. Parametric equations are given by $x = a(\theta - \sin(\theta))$, $y = a(1 - \cos(\theta))$, where a is the circle's radius. A wider class of curves can be created by entering the formulas $x = a_1\theta - b\sin(d\theta)$, $y = a_2 - c\cos(e\theta)$ in Cells F12..G12 of the basic parametric model and copying them.

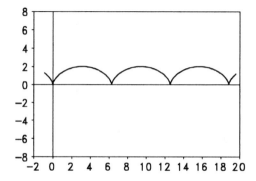

Figure 2.78. Cycloid: $x = \theta - \sin(\theta)$, $y = 1 - \cos(\theta)$.

2.13.A. Cycloid via Circles

One way to display a cycloid is to show the progress of a point P on a circle of radius a that rolls on the x-axis (Lockwood, 1961, pp. 80 to 81). The model of Figure 2.81 constructs a series of circles that show the location of the rolling circle of radius $a = 1$ at every 20° of rotation (Figure 2.79). Initially P is at the bottom of the circle at (0,0), so that its position angle is $-\pi/2$. The images also display the radius drawn from the center of the circle to the point P. The successive locations of P are connected to generate the cycloid.

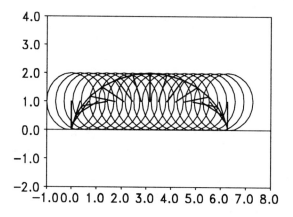

Figure 2.79. Cycloid via Rolling Circle.

The radius a is entered in Cell G3. Cell G9 gives the number of radians α in 20°. Rows 12..505 create the series of circles of 24 points produced via the standard multiple circle technique (see Section 2.4) using increments of 15° (Cell D3). Because the center (x,y) of the n-th circle is moved by $n\alpha a$ units in the x-direction while the y-coordinate is a, points on the n-th circle are created through the equations $x = n\alpha a + a\cos(\theta)$, $y = a + a\sin(\theta)$, for $0 \le \theta \le 2\pi$ as shown in Figure 2.80, with t representing α.

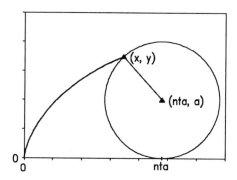

Figure 2.80. Rolling Circle.

Next, Rows 507..563 create the lines for the radii using the multiple line technique of Section 1.6. For the n-th circle, P has progressed counterclockwise around the circle by an angle $n\alpha$, to the relative position on the circle of $(a\cos(-\pi/2-n\alpha), a\sin(-\pi/2-n\alpha))$. Thus, the alternate points at the end of radii are

$$x = na\alpha+a\cos(-\pi/2-n\alpha) = na\alpha-a\sin(n\alpha), \quad y = a+a\sin(-\pi/2-n\alpha) = a-a\cos(n\alpha)$$

and the current location of the center, $x = na\alpha$, $y = a$. The block B507..H563 is named S.

The endpoints in S are listed so that the points on the circle are generated first for use with lookup functions. In Rows 565..583 the coordinates of successive points of the cycloid are generated by using lookup functions to find the points in S. The indicated formulas are entered in Row 565 and copied down through Row 583.

```
       B    C      D      E     F      G      H      I
   1 CYCLOID via CIRCLES
   2 Overall Parameters:
   3 Step:θ0=   15    a =      1
   :
   9 pi/180=0.017    α = 0.349 rad
  10 Circles:
  11   n  cnt   θ:rad     x     y1     y2     y3
  12   0   0    0.0     1.00  1.00
  13   0   1    0.3     0.97  1.26
   :
  36   0  24    6.3     1.00  1.00
  37   0  25    NA       NA    NA
  38   1   0    0.0     1.35  1.00
  39   1   1    0.3     1.31  1.26
   :
 505  18  25    NA       NA    NA
 506 Radii:
 507   0   0            0.00         0.00
 508   0   1            0.00         1.00
 509   0   2             NA           NA
 510   1   0            0.01         0.06
   :
 563  18   2             NA           NA
 564 Cycloid:
 565   0                0.00         0.00
 566   1                0.01         0.06
 567   2                0.06         0.23
   :
 583  18                6.28         0.00

D9:   @PI/180                G9:   20*D9
B12:  0                      B13:  +B12+(C13=0)
C12:  0                      C13:  @MOD(1+C12,26)
D12:  @IF(C12=25,@NA,15*D$9*C12)
F12:  +G$3*(G$9*B12+@COS(D12))
G12:  +G$3*(1+@SIN(D12))
B507: 0                      B508: +B507+(C508=0)
C507: 0                      C508: @MOD(1+C507,3)
F507: @CHOOSE(C507,B507*G$9*G$3-G$3*@SIN(G$9*B507),
      B507*G$9*G$3,@NA)
H507: @CHOOSE(C507,G$3*(1-@COS(B507*G$9)),G$3,@NA)
B565: 0                      B566: 1+B565
F565: @VLOOKUP(B565,$S,4)    I565: @VLOOKUP(B565,$S,6)
```

Figure 2.81. Cycloid via Rolling Circle.

70 Cycloids and Related Curves

Copy:		Graph:			
From	To	Series	Cells	Labels	Purpose
B13..C13	B14..C505	X	F12..F583		x-axis
D12..G12	D13..G505	1	G12..G505		circles
B508..C508	B509..C563	2	H12..H563		radii
F507..H507	F508..H563	3	I12..I583		cycloid
B566	B567..B583				
F565..I565	F566..I583				

2.13.B. Epicycloids

If n is a positive integer, and a circle of radius a rolls on the outside of the circumference of a circle of radius $(n-1)a$, then the path traced by a point P on the circumference of the rolling circle is called an *epicycloid*. The path is given by

$$x = na\cos(\theta) - a\cos(n\theta), \quad y = na\sin(\theta) - a\sin(n\theta)$$

The model of Figure 2.83 generates a more general class of curves, the *epitrochoids*, using

$$x = na\cos(\theta) - h\cos(n\theta), \quad y = na\sin(\theta) - h\sin(n\theta)$$

For $h = a$, this gives an epicycloid, as in Figure 2.82 with $n = 6$.

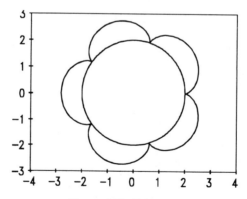

Figure 2.82. Epicycloid.

The radius r of the fixed circle is set in Cell D4, and n entered in Cell D6. The radius of the rolling circle is then calculated in Cell F6 as $a = r/(n-1)$ by +D4/(D6-1). To produce an epicycloid, set $h = a$ (+F6) in Cell F7. Cell F3 is used to rotate the epicycloid through any angle α, in degrees. The radian measure of α is computed in Cell F4. The indicated formulas are then placed in Row 12 and copied through Row 84 to produce the epicycloid. The fixed circle is generated in Rows 86..158.

```
         B      C       D       E       F      G       H
    1 PARAMETRIC EQUATIONS: Epicycloid
    2 Parameters:        Rotate:
    3 Step:θ =        5  α:deg=      0
    4 Circ:r =        2  α:rad=      0
    5 x=nacos(θ)-hcos(nθ),  y=nasin(θ)-hsin(nθ)
    6         n =    6        a =   0.4
    7                         h =   0.4
    :
   11   n   θ:deg  θ:rad          x     y1      y2
   12   0       0  0.000       2.00   0.00
   13   1       5  0.087       2.04   0.01
   14   2      10  0.175       2.16   0.07
    :
   84  72     360  6.283       2.00   0.00
   85 Circle:
   86   0       0  0.000       2.00           0.00
   87   1       5  0.087       1.99           0.17
    :
  158  72     360  6.283       2.00           0.00

  F4:  +D$9*F3    F6: +D4/(D6-1)    F7: +F6    D9: @PI/180
  B12: 0   B13: 1+B12   C12: +D$3*B12   D12: +D$9*C12
  F12: +D$6*F$6*@COS(D12+F$4)-F$7*@COS(D$6*D12+F$4)
  G12: +D$6*F$6*@SIN(D12+F$4)-F$7*@SIN(D$6*D12+F$4)
  B86: +B12   C86: +C12
  D86: +D12   F86: +D$4*@COS(D86)   H86: +D$4*@SIN(D86)
```

Figure 2.83. Epicycloid.

Copy:		Graph:			
From	To	Series	Cells	Labels	Purpose
B13	B14..B84	X	F12..158		x-axis
C12..G12	C13..G84	1	G12..84		epicycloid
B86..H86	B87..H158	2	H12..158		circle

2.13.C. Additional Curves

Several other curves related to the cycloid are listed below. Their constructions are left as activities. These curves are discussed further in Chapter 4.

1. *Prolate Cycloid*: $x = a\theta - b\sin(\theta)$, $y = a - b\cos(\theta)$, $a < b$. This is the path traced by a point P external to a circle of radius a, and b units from its center. As the circle rolls along a line, P moves so that it remains on the extension of a given radius (Figure 2.84).

2. *Curtate Cycloid*: $x = a\theta - b\sin(\theta)$, $y = a - b\cos(\theta)$, $a > b$. This is the path of a point internal to a circle of radius a as it rolls along a line. P is b units from the center (Figure 2.85).

3. *Hypocycloid*: $x = na\cos(\theta) + a\cos(n\theta)$, $y = na\sin(\theta) - a\sin(n\theta)$. This is the path of a point on a circle of radius a that rolls on the inner side of the circumference of a circle of radius $(n+1)a$ (Figure 2.86).

4. *Epitrochoid*: $x = na\cos(\theta) - h\cos(n\theta)$, $y = na\sin(\theta) - h\sin(n\theta)$. This is the path of an point P external to a circle of radius a that rolls on the outer circumference of the circle. P is h units from the center of the circle (Figure 2.87).

72 Cycloids and Related Curves

5. *Hypotrochoid*: $x = na\cos(\theta) + h\cos(n\theta)$, $y = na\sin(\theta) - h\sin(n\theta)$. This is the path of a point P internal to a circle of radius a that rolls on the inside of the circumference of the circle. P is h units from the center of the circle (Figure 2.88).

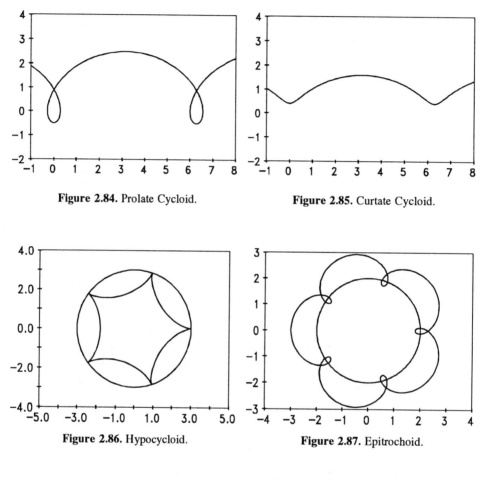

Figure 2.84. Prolate Cycloid.

Figure 2.85. Curtate Cycloid.

Figure 2.86. Hypocycloid.

Figure 2.87. Epitrochoid.

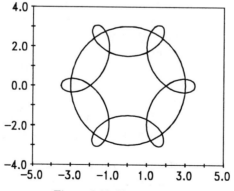

Figure 2.88. Hypotrochoid.

Activities

1. Create models for each class of curves in Part C. Often, more than one of the curves can be formed from the same model. Experiment with many values of the parameters.
2. Create two epicycloids or hypocycloids within one graph, as in Figures 2.89 and 2.90.

Figure 2.89. Hypocycloids.

Figure 2.90. Hypocycloid and Epicycloid.

3. Create and investigate curves that are produced by other curves as the roll along a variety of other curves. This topic is presented and analyzed in Hall and Wagon (1992).

2.14. A MÉLANGE OF CURVES

Many other curves can be created using the basic model of Section 2.2. Some are listed below. Additional curves are found in von Seggern, 1993, Lawrence, 1972, and other references. In designing a model, it is useful to employ parameters and to make the model more universal. Additional aspects of the first three curves are discussed in Chapter 3.

1. *Cissoid*: $r = a(\csc(b\theta) - \sin(c\theta))$.
2. *Involute of Circle*: $x = a(\cos(\theta) + \theta\sin(\theta))$, $y = a(\sin(\theta) - \theta\cos(\theta))$
3. *Conchoid*: $r = a\csc(b\theta) + c$.
4. *Witch of Agnesi*: $x = 2a\cot(\theta)$, $y = 2a\sin^2(\theta)$.
5. *Piriform*: $x = a(1+\sin(\theta))$, $y = b(1+\sin(\theta))\cos(\theta)$.
6. *Tschirnhausen's Cubic*: $r = a\sec^3(\theta/3)$.

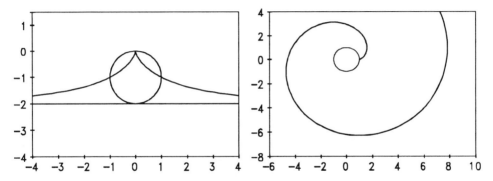

Figure 2.91. Cissoid, $r = -2(\csc(\theta) - \sin(\theta))$.   **Figure 2.92.** Involute of Circle.

74   A Mélange of Curves

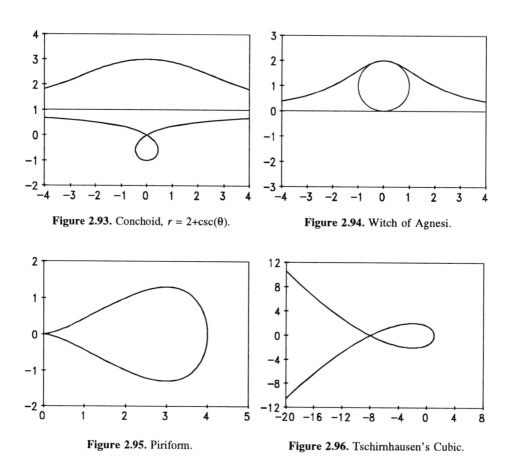

Figure 2.93. Conchoid, $r = 2+\csc(\theta)$.

Figure 2.94. Witch of Agnesi.

Figure 2.95. Piriform.

Figure 2.96. Tschirnhausen's Cubic.

Activities

1. Create models that include parameters for each of the curves listed above. Study the effects of changing the values of the parameters.
2. Examine the references for more sources of curves to implement, including the *bifolium*, $r = a\sin(b\theta)\cos^2(b\theta)$, and the *semicubical paraboloid*, $r = (\tan^2(\theta)\sec(\theta))/a$.
3. Some mathematical algorithms can be used to generate beautiful drawings. The relations below produce the rose drawings discussed in Maurer (1987) as Method A. Here, rad$\{\theta\}$ indicates converting θ from degrees to radians.

$$\theta_0 = 0, \ \theta_{i+1} = \theta_i + d \ (\text{mod } 360)$$

$$s_i = \text{rad}\{n\theta_i \ (\text{mod } 360)\}, \ r_i = \sin(s_i)$$

$$t_i = \text{rad}\{\theta_i\}, \ x_i = r_i\cos(t_i), \ y_i = r_i\sin(t_i)$$

Create a model for this algorithm, and investigate curves that are produced by varying the parameters n and d. Figure 2.97 provides an outline. Figures 2.98 and 2.99 use the values $n = 8$, $d = 308.5$ and $n = 2.5$, $d = 97$, respectively. Try to implement Maurer's Method B as well. Other algorithms for similar rosettes are discussed in Hill (1990), pp. 107 to 110.

```
      A       B       C       D       E       F       G
1  ROSE    n =          8   d =   308.5
2
3     n       θ       s       r       t       x       y
4     0     0.0   0.000   0.000   0.000   0.000   0.000
5     1   308.5   5.376  -0.788   5.384  -0.491   0.617
6     2   257.0   4.468  -0.970   4.485   0.218   0.945

A4: 0   A5: 1+A4        B4: 0   B5: @MOD(B4+E$1,360)
C4: @RADIANS(@MOD(B4*C$1,360))
D4: @SIN(C4)            E4: @RADIANS(B4)
F4: +D4*@COS(E4)        G4: +D4*@SIN(E4)
```

Figure 2.97. Rose Model.

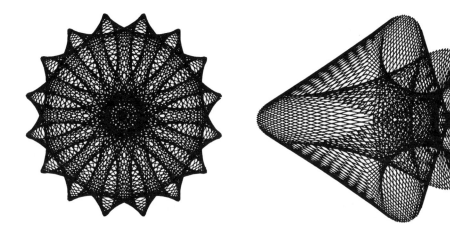

Figure 2.98. Rose: *n* = 8, *d* = 308.5.　　　**Figure 2.99.** Rose: *n* = 2.5, *d* = 97.

2.15. SKETCHING WITH SPREADSHEET SOLVER COMMANDS

The (x,y)-coordinates of each of the curves previously considered ultimately were described by functions of a single parameter, usually denoted by θ. The curves were created by generating the (x,y)-values for a sequence of equally spaced values of θ. However, some curves are not so easily described, and are defined by more complex equations in x and y. This section provides a brief introduction to techniques that use a spreadsheet's equation solver command in sketching such curves. Although a full analysis of these curves often requires a more detailed study (e.g., Hilton, 1932; Frost, 1960; Salmon, 1960), a spreadsheet still can be used effectively in producing their graphs. While not all spreadsheets possess equation solvers, both *Quattro Pro* and *Excel* do. Others, including *Lotus 1-2-3*, can incorporate them through add-on programs.

The basic ideas of the use of a spreadsheet's equation solver can be illustrated through the example of Figure 2.100. Here, a solution (x,y) of the equation $xy^2 - x + y^3 = 3$ is desired so that direction from the origin to (x,y) is given by a selected polar angle θ. The value of θ is placed in Cell C2, and an estimate, r_0, for the distance of the point (x,y) from the origin is entered in Cell D2. Next, r, the absolute value of r_0, is computed in Cell E2 as @ABS(D2). This enables the r of the polar equations for x and y to be restricted to positive values, an aspect that is mentioned later. The x- and y-components are determined from r and θ as $x = r\cos(\theta)$ and $y = r\sin(\theta)$ in Cells F2...G2. Finally, the value of $f(x,y) = xy^2-x+y^3$ is calculated in Cell H2. Using $\theta = 0.20$ radians and $r_0 = 5.00$, the corresponding values that are produced are $(x,y) = (4.90, 0.99)$, with $f(x,y) = 0.92$. Clearly, the estimate for r_0 was incorrect, because $f(x,y) \neq 3$.

```
       C      D      E      F      G      H
1      θ      r0     r      x      y      fn
2      0.20   5.00   5.00   4.90   0.99   0.92

C2: 0.2     D2: 5           E2: @ABS(D2)
F2: +E2*@COS(C2)            G2: +E2*@SIN(C2)
H2: +F2*G2^2-F2+G2^3
```

Figure 2.100. Basic Solver Example (Before).

A user can systematically experiment with manual trial and error substitution for r_0 in Cell D2 to find increasingly better approximations of the desired solution. As the value of Cell D2 is altered, Cell H2 also is changed. However, the spreadsheet solver command can carry out this process automatically. To use the solver, appropriate menu selections are made so that Cell H2 is selected as the location of the formula to be set to a certain target value, 3 is set as the target value, and Cell D2 is chosen as the variable cell to change to arrive at the desired target. The degree of accuracy of the approximation, for example, to within 0.001 of the target value, and the maximum number of iterations that will be allowed, are also assigned. When this is done, issuing a "Go" command causes the solver to carry out the approximation process. The solver iteratively adjusts Cell D2 until the value of Cell H2 approximates the set target value to within the selected degree of accuracy, or until the number of iterations allowed is reached. If the latter number is exceeded, a message appears on the screen. Figure 2.101 shows the result of using the solver. It indicates that the point $(x,y) = (5.58, 1.13)$ produces $f(x,y) = 3$. Thus, this point satisfies the given equation.

```
       C      D      E      F      G      H
1      θ      r0     r      x      y      fn
2      0.20   5.69   5.69   5.58   1.13   3.00
```

Figure 2.101. Basic Solver Example (After).

The specific solver commands from *Quattro Pro* that are used for this example are listed below.

```
/ T(ool) S(olve for)
  F(ormula cell): H2   T(arget Value): 3   V(ariable Cell): D2
  P(arameters) M(aximum iterations): 99  A(ccuracy): 0.001
  G(o)  Q(uit)
```

The commands for *Excel* are similar.

> Formulas Solver
> Set Cell: **H2** Equal To: **3** By Changing: **D2**
> Solve

In using a solver, there are several things to consider. As usual, there are some differences among the various spreadsheets. For example, *Excel* allows more than one cell to be adjusted, and permits a user to choose among various types of approximation algorithms. Second, a particular choice of r_0 may produce a very slow rate of convergence to a solution, or it may not result in convergence at all. In the first case, one can either repeat the process by resuming from the value last computed, or manually alter the initial approximation and repeat.

This section shows three settings in which a spreadsheet solver command can be employed in graphing curves defined by equations. Explorations into the concepts related to other types of equations can be pursued as well by using the books listed above.

2.15.A. Lemniscates

Section 2.9 describes a special type of lemniscate. The model below examines a wider class of lemniscates using a more general definition (Markushevich, 1980). Suppose that a positive number k and a set of n points, (a_i,b_i), $i = 1, 2,\ldots, n$ are given. The set of all points (x,y) such that the product of the distances from (x,y) to the points (a_i,b_i) is equal to k is called a *lemniscate with n foci*. An equation satisfied by such points is given by

$$((x-a_1)^2+(y-b_1)^2)((x-a_2)^2+(y-b_2)^2)\ldots((x-a_n)^2+(y-b_n)^2) = k^2$$

With the choices of $n = 2$, $k = 1/2$, and foci of $(\pm1/\sqrt{2},0)$, the lemniscate of Bernoulli of Figure 2.55 will be produced.

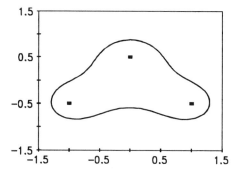

Figure 2.102. Lemniscate (three foci), $k = 1.1$.

For some values of k a single curve is produced, while for others, multiple closed curves result. The model of Figure 2.103 is designed to create either the single curve or one of the multiple curves that is created from three foci. The model uses the solver to find solutions of $f(x,y) = k^2$, where

$$f(x,y) = ((x-a_1)^2+(y-b_1)^2)((x-a_2)^2+(y-b_2)^2)((x-a_3)^2+(y-b_3)^2)$$

The lemniscate of Figure 2.102 has three foci: (-1.0,-0.5), (0.0,0.5), (1.0,-0.5), and $k = 1.1$.

To create the model, the foci (a_i,b_i) are entered as parameters in Cells C3..D5. The constant k is set in Cell G3, with k^2 calculated in Cell G4 as + G3^2. The underlying idea of this model is to plot points on the curve that surround a chosen center point (a,b) that is located in the "middle" of the foci, and will lie in the interior of the curve that is produced. The coordinates of (a,b) are set in Cells C6...D6. These can be entered either as constants or created by formulas such as C6: @AVG(C3..C5) and D6: @AVG(D3..D5). The model plots 72 points proceeding counterclockwise around the center (a,b) in steps of 5°. Here, θ_0, the radian equivalent of 5°, is created in Cell G5 as @PI/36.

Rows 12..84 are used to generate the (x,y)-coordinates of the points around the center (a,b). Column B provides a point counter. Column C then generates angles θ in steps of θ_0 radians. Next, a user provides an estimate of the positive distance r_0 from center to a typical point on the lemniscate, enters it in Cell D12, and then copies that value down Column D.

The values of r_0 will be changed subsequently by the solver. Because there may be more than one solution to the equation for a given target value, the model selects one with a positive distance r. To accomplish this, $r = |r_0|$ is calculated in Column E using the library function @ABS.

Next, the model computes the (x,y)-coordinates in Columns F..G from the polar coordinates r,θ, as $x = a + r\cos(\theta)$, $y = b + r\sin(\theta)$. Finally, values of the function $f(x,y)$ are calculated in Column H. To create the table, the indicated formulas are entered in Row 12 and copied through Row 84 to produce very rough estimates of points on the curve. The graph of the resulting curve is created by using Columns F..G to provide the respective x- and y-values. Initially, using only the initial crude estimates, this will generate a circle.

	B	C	D	E	F	G	H
1	Lemniscate:						
2	Foci:	x	y		Constant:		
3	Pt 1	-1	-0.5		k =	1.1	
4	Pt 2	0	0.5		k² =	1.21	
5	Pt 3	1	-0.5		pi/36=	0.09	
6	Cntr:	0	0				
:							
11	n	θ:rad	r0	r	x	y1	fn
12	0	0.00	1.00	1.00	1.00	0.00	1.33
13	1	0.09	1.00	1.00	1.00	0.09	1.74
14	2	0.18	1.00	1.00	0.98	0.17	2.15
:							
83	71	6.20	1.00	1.00	1.00	-0.09	0.95
84	72	6.28	1.00	1.00	1.00	0.00	1.33

```
G4:  +G3^2    G5: @PI/36    B12: 0     B13: 1+B12
C12: +B12*G$5+0.001         D12: 1     E12: @ABS(D12)
F12: +C$6+E12*@COS(C12)     G12: +D$6+E12*@SIN(C12)
H12: ((F12-C$3)^2+(G12-D$3)^2)*
     ((F12-C$4)^2+(G12-D$4)^2)*
     ((F12-C$5)^2+(G12-D$5)^2)
```

Figure 2.103. Lemniscate of Three Foci (Before).

Copy:		Graph:			
From	To	Series	Cells	Labels	Purpose
B13 C12..H12	B14..B84 C13..H84	X 1	F12..F84 G12..G84		x-axis curve

However, as described above, the equation solver now is used to adjust each of the rough estimates so that the (x,y)-points will solve the equation. First, the initial values in Row 12 are adjusted, just as in the basic example above, by using Cell H12 as the formula cell, Cell D12 as the variable, and the value of Cell G4 (here, 1.21) as the target value. When the solver is invoked, r_0 is changed from 1 to 0.96 to give $x = 0.96$, $y = 0.00$, and $f(x,y) = 1.21$ as desired. Thus, the point $(x,y) = (0.96, 0.00)$ lies on the desired curve.

This process is repeated, next using D13 and H13 as the variable and formula cells, and so on through Row 84 to produce the desired points on the curve. Some of the resulting numerical output is shown in Figure 2.104, with the graph that is generated displayed in Figure 2.102. The model has been supplemented to show the foci in the graph. As in earlier examples, the insert and copy procedure has been used (with Cell G5 adjusted accordingly) to generate 72 additional points and to form a smoother curve.

```
        B     C      D      E      F      G      H
11      n    θ:rad   r0     r      x      y1     fn
12      0    0.00    0.96   0.96   0.96   0.00   1.21
13      1    0.09    0.86   0.86   0.85   0.08   1.21
14      2    0.18    0.79   0.79   0.78   0.14   1.21
15      3    0.26    0.76   0.76   0.73   0.20   1.21
16      4    0.35    0.73   0.73   0.69   0.25   1.21
 :
83     71    6.20    1.10   1.10   1.09  -0.09   1.21
84     72    6.28    0.96   0.96   0.96   0.00   1.21
```

Figure 2.104. Lemniscate of Three Foci (After).

Naturally, this model possesses neither the full flexibility nor the degree of interaction of previous models. To see the effects of any changes in the foci or in the value of k, the entire solver process must be repeated. This process can take a while to complete. Fortunately, the operation can be accelerated and streamlined through the use of macros, which are spreadsheet programs. Spreadsheet instruction books and manuals should be consulted for detail on their design and use. However, one useful macro in *Quattro Pro* is described in Figure 2.105.

```
        O       P                       R
1       N =                             Loop counter
2
3       \G   '/tstg4~q                  Set target value: Cell G4
4            '/tsfH12~vD12~gq           Set formula cell:H12, variable cell:D12
5            {FOR P1,1,72,1,P7}         Loop: FOR P1 = 1 TO 72 STEP 1  DO P7
6
7            '/tsf{down}~v{down}~gq     Move formula, target cells down 1
```

Figure 2.105. Solver Macro.

80   *Sketching with Spreadsheet Solver Commands*

One way to create a macro is to enter the codes of a series of commands in a column of adjacent cells of the spreadsheet model, as in P3..P5 and give it a name starting with \. Here, the range P3..P5 is named \G. The <ALT>G keys are pressed to invoke the macro of Figure 2.105. Macros also can be created in separate files or by recording the steps as they are carried out by hand. The macro expressions in Cells P3, P4, and P7 represent a series of *Quattro Pro* keystrokes that carry out solver commands. Cell P5 is used to implement a FOR loop, carrying out the macro command in Cell P7 a total of 72 times.

This model can also be used to create a great variety of lemniscates, including those that have a repeated focus. Figure 2.106 shows the lemniscate with three foci that is produced by using the left point twice as a focus. The foci are (-1,0), (-1,0), and (1,0), with $k = 1.3$.

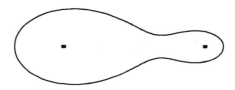

Figure 2.106. Lemniscate: Repeated Left Focus.

It is also straightforward to modify the model for other numbers of foci. The lemniscates of Figures 2.107 and 2.108 have four and five foci, respectively.

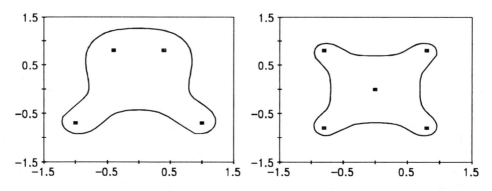

Figure 2.107. Lemniscate with Four Foci.   **Figure 2.108.** Lemniscate with Five Foci.

Difficulties can arise in using a model constructed along these lines. It may be hard to locate a center, a value of k that encompasses the foci, or a good value of r_0. If the choice of k produces a multiple-component curve, the curve sketching process may jump from one component to another. In this case, the process must be redone using alternate values of k or r_0. Also, using only a single initial choice for r_0 for all points may mean that the computations for some points take longer to produce convergence. Moreover, the number of steps required for convergence at a given point may exceed the number of calculations allowed. In this case the solver command can be used continually to recalculate the point, but doing so is awkward when the difficulty occurs in the midst of a macro calculation.

2.15.B. Higher Degree Plane Algebraic Curves

Frequently the graph of an equation consists of two or more disjoint component curves. The next model modifies the model of Section A to create the graph of an equation that has two component curves. A center point is selected for each part. Although it is easy to design the modifications in the model, in practice it may be difficult to know how many components to search for, or where the centers should be located. Books, such as those listed at the start of this section, on advanced curve sketching techniques should be consulted. As an illustrative example, Figure 2.109 shows the graph of the equation $(5y+x^2)^2-4x(x-3)(x+5) = 0$ that is created by the model of Figure 2.110. (See Hilton, 1932, p. 331.) The graph consists of a closed curve together with an open-ended branch that extends infinitely at each end.

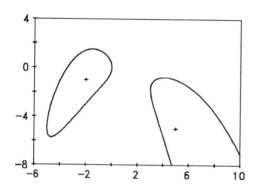

Figure 2.109. $(5y+x^2)^2-4x(x-3)(x+5) = 0$.

```
         B         C        D       E        F         G        H
  2  Center:       x        y       α   Constants:
  3    Pt1        -2       -1       0   Targ=         0
  4    Pt2         5       -5    -0.5   pi/36      0.09
  :
 11         n   θ:rad       r      r1        x        y1       fn
 12         0    0.00    1.50    1.50    -0.50     -1.00    -8.94
 13         1    0.09    1.50    1.50    -0.51     -0.87   -15.14
 14         2    0.18    1.50    1.50    -0.52     -0.74   -21.26
  :
 84        72    6.28    1.50    1.50    -0.50     -1.00    -8.94
 85  Second:                                NA        NA
 86         0   -0.50    2.00    2.00     6.76     -5.96   -941.9
 87         1   -0.41    2.00    2.00     6.83     -5.80   -927.0
 88         2   -0.33    2.00    2.00     6.90     -5.64   -903.6
  :
158        72    5.78    2.00    2.00     6.76     -5.96   -941.9

G4:  @PI/36
B12: 0       B13: 1+B12        C12: +E3      C13: +C12+G$4
D12: 1.5                       E12: @ABS(D12)
F12: +C$3+E12*@COS(C12)   G12: +D$3+E12*@SIN(C12)
H12: (5*G12+F12^2)^2-4*F12*(F12-3)*(F12+5)
B86: +B12                      C86: +E4      C87: +C86+G$4
D86: 2                         E86: @ABS(D86)
F86: +C$4+E86*@COS(C86)   G86: +D$4+E86*@SIN(C86)
H86: (5*G86+F86^2)^2-4*F86*(F86-3)*(F86+5)
```

Figure 2.110. Graph with Two Components (Before).

Copy:		Graph:			
From	To	Series	Cells	Labels	Purpose
B13..C13	B14..C84	X	F12..F158		x-axis
D12..H12	C13..H84	1	G12..G158		curve
B86	B87..B158				
C87	C88..C158				
D86..H158	D87..H158				

In Figure 2.110 a center, (-2,-1), for the closed segment is entered in Cells C3..D3. Because the graphs of equations will not always form closed curves, one can choose the direction in which to start plotting points by an angle $\theta = \alpha$, whose value is set in Cell E3. Here, α is set to 0. The center for the second component is set similarly in Cells C4..D4, with α set in Cell E4. Experimentation shows that the value $\alpha = -0.5$ will start the right branch at a convenient point.

The first curve (the closed left loop) is created in Rows 12..84 essentially as in Section A. The values for θ are created slightly differently in Column C to allow for the initial value of α. Also, the function in Column H is adjusted for the current equation. An initial estimate of $r_0 = 1.5$ has been entered into Cells D12..D84. Now the solver is used as before, after setting the target value to 0. The results can be seen in the upper part of Figure 2.111.

The second component curve is created similarly in Rows 86..158 with formulas that employ the second center of Row 4. This time an estimate of $r_0 = 2.0$ is entered in Cells D86..D158. Similar steps can be used with the solver, or a second macro can be created as a slight modification of the first to update the estimates. Because the second branch has values that go to infinity, eventually the solver routine will fail to converge. To keep from continually adjusting the estimate for points that lie far from the center, the process can be stopped and @NA entered for the x- and y-values in this range. The output that results from using the solver is displayed in Figure 2.111.

	B	C	D	E	F	G	H
11	n	θ:rad	r	r1	x	y1	fn
12	0	0.00	1.62	1.62	-0.38	-1.00	-0.00
13	1	0.09	1.73	1.73	-0.28	-0.85	-0.00
14	2	0.18	1.84	1.84	-0.18	-0.68	-0.00
⋮							
84	72	6.28	1.62	1.62	-0.38	-1.00	-0.00
85	Second:				NA	NA	
86	0	-0.50	-6.22	6.22	10.46	-7.98	0.0
87	1	-0.41	5.56	5.56	10.10	-7.23	0.0
88	2	-0.33	-5.05	5.05	9.78	-6.61	0.0
⋮							
147	61	4.82	6.94	6.94	5.77	-11.90	0.0
148	62	4.91	-0.80	0.80	NA	NA	NA
⋮							
158	72	5.78	2.00	2.00	NA	NA	NA

Figure 2.111. Graph with Two Components (After).

The variety of curves that can be produced this way is quite extensive. Two more illustrative examples are provided below. The equation that generates Figure 2.112 is

$((y^2+x^2)^2-4a^2x^2)(x^2+y^2-4a^2) = 0$ with a = 1 (Frost, 1960, p. 200), while that of Figure 2.113 is $9(x+4)y^2+24xy+x(x^2+2x-16) = 0$ (Hilton, 1932, p. 214).

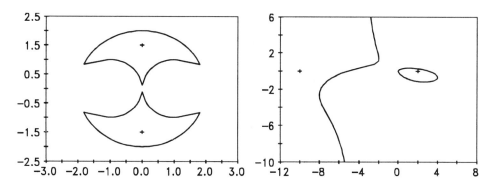

Figure 2.112. Multiple Components I.   Figure 2.113. Multiple Components II.

2.15.C. Taxicab Geometry

The distance d_E between two points (x_0,y_0) and (x_1,y_1) in Euclidean geometry, the geometry that is used throughout this book, is given by

$$d_E = \sqrt{(x_0-x_1)^2+(y_0-y_1)^2}$$

However, this is not the only possible definition of distance. In *taxicab geometry* (Krause, 1986) the distance between two points is defined by

$$d_T = |x_0-x_1| + |y_0-y_1|$$

This definition corresponds to measuring distances as the sum of the distances obtained by moving only horizontally and vertically between points, much as a taxicab would travel on a rectangular city street grid. Using the spreadsheet solver techniques above, it is possible to generate taxicab conic curves, as shown in Figures 2.114 to 2.116. It should be noted that in investigating these curves some cases give rise to anomalies that are not encountered in Euclidean geometry, and some care may be needed in the size of θ_0 to ensure that sharp corners are produced.

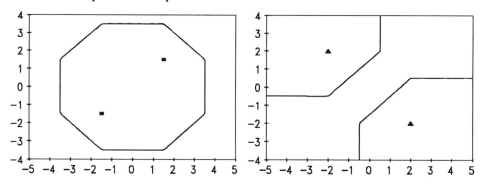

Figure 2.114. Taxicab Ellipse.   Figure 2.115. Taxicab Hyperbola.

84 *Sketching with Spreadsheet Solver Commands*

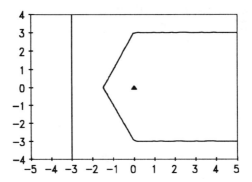

Figure 2.116. Taxicab Parabola.

Activities

1. Modify the model of Figure 2.103 for lemniscates with other than three foci. Experiment with the location of foci, the choice of the graphing center, and the value of k.
2. Use the model of Part B to create lemniscates having two components, as in Figure 2.117. Use a value for k less than half the distance between the foci. Also use this approach to create the lemniscate of Bernoulli (see Part A above).

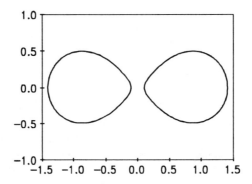

Figure 2.117. Lemniscate with Two Foci.

3. Examine Salmon (1960), Frost (1960), Yates (1974), and Hilton (1932) for plane curves. Draw the *bicorn (cocked hat)* $x^2(x^2+y^2)+4ax^2y-2a^2x^2+3a^2y^2-4a^3y+a^4 = 0$.
4. Create graphs of the curves produced by the following equations from Teixeira (1971). Modify the models for use with polar equations where needed.
   (a) $x^4 - ax^3 + b^2y^2 = 0$ (I, pp. 289 to 290); (b) $(x^2+y^2) = ax^3$ (I, p. 297); (c) $r^4 - 2ar^3\cos(3\theta) + a^2r^2 - a^4 = 0$ (III, p.53).
5. Create the following curves and others from pp. 71 to 75 of Cundy and Rollett (1961). (a) $y^2 = x^4 + x^5$; (b) $(x^2-1) = y^2(y-1)(y-2)(y+5)$ *(stirrup)*; (c) $xy(x^2-y^2) = x^2+y^2$ *(Maltese cross)*; (d) $y^2(y^2-96) = x^2(x^2-100)$ *(electric motor)*; (e) $y^4-x^4 = xy$; (f) $y^2 = x^4 - x^6$.
6. Create models that produce the taxicab curves of Part C above. Also create taxicab "circles" and "perpendicular" bisectors as discussed in Krause (1986).

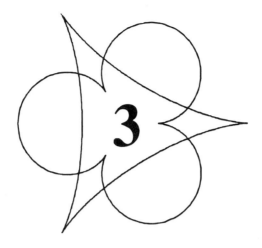

Chapter 3

CONSTRUCTIONS

The examples of the previous chapter were designed to sketch specific curves as they are defined by equations. In comparison, the models in this chapter are formulated to develop classes of curves defined through geometric constructions. Each section contains the definition of a particular class of curves and presents a model for its construction. Although each example is implemented with a specific base curve, that curve can be changed to generate other designs that are based on the same geometric construction. Discussions of most of the constructions are contained in Lockwood (1961), Teixeira (1971), and Yates (1974).

The descriptive figures that are shown to illustrate the definitions have been produced through slightly expanded versions of the same models. Thus, these examples also illustrate how graphs produced by a spreadsheet can be developed into interactive geometric illustrations that can be incorporated into both written documents and visual presentations.

3.1. STROPHOIDS

Let C be a curve, O a point called a pole, and A a fixed point as shown in Figure 3.1. Let L be a line drawn through O so that it intersects the curve C at the point Q, and let P_1 and P_2 denote points on L for which $P_1Q = QP_2 = QA$. The locus of the points P_1 and P_2 as Q varies over the curve C is the *strophoid of C with respect to A and O*. Details are contained in Lockwood (1961), pp. 90 to 97; Yates (1974), pp. 217 to 220; Teixeira (1971), pp. 30 to 45 of Tome I; and Lawrence (1972), pp. 100 to 104. The name strophoid is derived from "twisted strap".

The model of Figure 3.2 generates the circle of radius a and center (0,0) for the curve C as a special case of $r = a + b\cos(\theta)$. The values of a and b are set in Cells D6 and F6, with the coordinates of the pole $O = (x_0, y_0)$ entered in Cells F3..F4, and those of $A = (u_0, v_0)$ put in Cells H3..H4. The base curve C is constructed in Cells B12..G84 using the standard procedure of Chapter 2. The incremental step size, θ_0, for the angle θ is set in Cell D3. To derive the other formulas in Row 12, consider a typical point $Q = (x_1,y_1)$ of C. The coordinates of the first point Q of C are given in Cells F12 and G12. Cell I12 determines k, the distance QA, as

$$k = \sqrt{(x_1-u_0)^2+(y_1-v_0)^2}$$

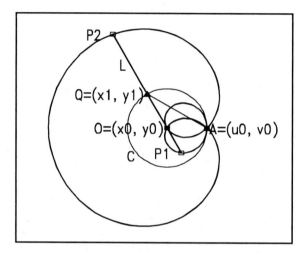

Figure 3.1. Strophoid (Base: Circle).

To determine points on the line passing through O and Q, first observe that the vector $[x_1-x_0, y_1-y_0]$ is in the correct direction. Cell H12 computes its length d as

$$d = \sqrt{(x_1-x_0)^2+(y_1-y_0)^2}$$

Thus, from Section 1.2, a vector in the required direction that has length k is given by

$$(k/d)[x_1-x_0, y_1-y_0]$$

and the desired points P_1 and P_2 are obtained as

$$(x_1,y_1) \pm (k/d)[x_1-x_0, y_1-y_0]$$

These points determine two curves. The curve determined by the minus sign is designated as the "lower" curve. Cells J12..K12 find the point $P_1 = (x_{lo}, y_{lo})$ on that curve as

$$x_{lo} = x_1 - (k/d)(x_1-x_0), \quad y_{lo} = y_1 - (k/d)(y_1-y_0)$$

The point P_2 on the "upper" curve is found similarly in Cells L12..M12 from the plus sign. The block is completed by copying the formulas in Row 12 down through Row 84 and naming the range B12..M84 as T. Thus, the coordinates of the lower curve appear in

Columns J..K and those of the upper curve in Columns L..M. To generate the graphs of these curves, the values for the lower curve are reproduced in Columns F and N of Rows 86..158 by entering +J12 and +K12 in Cells F86 and N86 and copying them through Row 158. The upper curve is formed similarly in Columns F and O of Rows 160..232.

```
        B   C     D     E     F     G     H    I    J    K    L    M    N     O
 1  STROPHOID (Cardioid/Circle)
 2  Parameters:   Pole:O    Fixed Pt:A
 3  Step:θ0=   5   x0=   0   u0=   1
 4                y0=   0   v0=   0
 5  Curve Parameters:  r = a + b·cos(θ)
 6       a=    1  b =   0
 :
 9  pi/180=0.017                                                     Curves
10  Curve:                    Circ                                   Lo   Hi
11   n  θ:d   θ:r   r     x    y1    d    k   Xlo  Ylo  Xhi  Yhi    y2   y3
12   0   0   0.00  1.0  1.00  0.00  1.0  0.0  1.0  0.0  1.0  0.0
13   1   5   0.09  1.0  1.00  0.09  1.0  0.1  0.9  0.1  1.1  0.1
14   2  10   0.17  1.0  0.98  0.17  1.0  0.2  0.8  0.1  1.2  0.2
 :
84  72 360   6.28  1.0  1.00  0.00  1.0  0.0  1.0  0.0  1.0  0.0
85  Lower Curve:
86   0              1.00                                              0.0
87   1              0.91                                              0.1
88   2              0.81                                              0.1
 :
158 72               1.00                                              0.0
159 Upper Curve:
160  0               1.00                                                   0.0
161  1               1.08                                                   0.1
162  2               1.16                                                   0.2
 :
232 72               1.00                                                   0.0

B12: 0             B13: 1+B12
C12: +D$3*B12      D12: +C12*D$9          E12: +D$6+F$6*@COS(D12)
F12: +E12*@COS(D12)                       G12: +E12*@SIN(D12)
H12: @SQRT((F12-F$3)^2+(G12-F$4)^2)
I12: @SQRT((F12-H$3)^2+(G12-H$4)^2)
J12: +F12-I12/H12*(F12-F$3)               K12: +G12-I12/H12*(G12-F$4)
L12: +F12+I12/H12*(F12-F$3)               M12: +G12+I12/H12*(G12-F$4)
B86: +B12    F86: +J12                    N86: +K12
B160: +B12   F160: +L12                   O160: +M12
```

Figure 3.2. Strophoid (Base: Circle).

Copy:		Graph:			
From	To	Series	Cells	Labels	Purpose
B13	B14..B84	X	F12..F232		x-axis
C12..M12	C13..M84	1	G12..G84		circle
B86..N86	B87..N158	2	N12..N158		lower curve
B160..O160	B161..O232	3	O12..O232		upper curve

The model can be augmented and modified to include the generating lines L in the graph, as in Figure 3.3. Here, the inner loops represent the lower curve, the outer loops the upper curve. The technique of Section 1.6 is used to create the line segments.

88 *Strophoids*

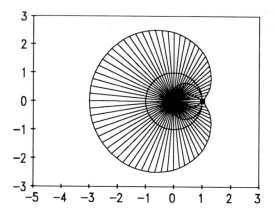

Figure 3.3. Strophoid (Base: Circle).

<u>Activities</u>

1. Examine the effects of changes in the locations of the fixed point and the pole. Figures 3.4 and 3.5 illustrate two new strophoids that are produced in this way.

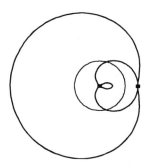 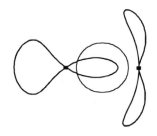

Figure 3.4. Strophoid of Circle. **Figure 3.5.** Strophoid of Circle.

2. Create strophoids using other base curves as in Figures 3.6 and 3.7. The base curve is not shown in Figure 3.7.

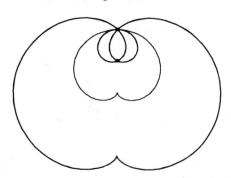 

Figure 3.6. Strophoid of Cardioid. **Figure 3.7.** Strophoid of Curve $r = \cos(2\theta)$.

3. Modify the model of Figure 3.2 to produce graphs that include construction lines and points.
4. For some choices of points and curves, either ERR or extremely large values may result from division by zero when $d = 0$, producing some anomalies in the graphs. Design formulas that overcome this difficulty.

3.2. CONCHOIDS

Let C be a curve, A a fixed point, and k a constant. Let Q be an arbitrary point on C, L the line through A and Q, and P_1 and P_2 two points on L so that $P_1Q = QP_2 = k$, as shown in Figure 3.8, where C is a line. The locus of the points P_1 and P_2 as Q varies over C is a *conchoid of C with respect to A*. This curve typically has two branches. Additional ideas are found in Lockwood (1961), pp. 126 to 129; Teixeira (1971), pp. 259 to 268 of Tome I; Yates (1974), pp. 31 to 35; Lawrence (1972), pp. 137 to 139; and Pedoe (1976), pp. 242 to 245. Conchoid means "shell-shaped".

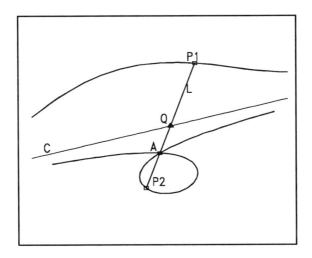

Figure 3.8. Conchoid (Base: Line).

The model of Figure 3.9 uses the line $y = ax+b$ for C. The values for a and b are set in Cells D6 and F6, with the coordinates of $A = (x_0, y_0)$ in Cells F3..F4, the constant k in Cell H3, and the initial x value in Cell D4.

The base curve C is created in Rows 12..84, with the points $Q = (x_1, y_1)$ that lie on C given in Columns F and G. The x-values are created in Column F by reproducing the initial x as +D4 in Cell E12 and creating subsequent x-values by using the incremental size h_0 (here, 0.5) given in Cell D3. The y-values on the line are formed in Column G as $ax+b$ by entering +D$6*F12+F$6 in Cell G12 and copying. Cell H12 determines the distance d from A to the initial point Q on C by

$$d = \sqrt{(x_1-x_0)^2+(y_1-y_0)^2}$$

Conchoids

The computations in the next columns are designed to construct points (x_2, y_2) on the conchoid. The vector from A to Q, $[x_1-x_0, y_1-y_0]$, has length is d. Thus, a unit vector in the same direction is given by $(1/d)[x_1-x_0, y_1-y_0]$, while a vector of length k is

$$(k/d)[x_1-x_0, y_1-y_0]$$

There are two points on the vector that are k units from Q. These points produce two branches of the conchoid. As in the previous section, points on the lower curve are located in Columns J and K with the first point determined in Cells J12..K12 by

$$(x_2, y_2) = (x_1, y_1) - (k/d)[x_1-x_0, y_1-y_0]$$

The upper curve is generated in Columns L and M in same way as

$$(x_3, y_3) = (x_1, y_1) + (k/d)[x_1-x_0, y_1-y_0]$$

with the initial point produced in Cells L12..M12. Finally, the entries in Row 12 are copied down through Row 84.

The values for the lower and upper curves are used to create their graphs in Rows 86..158, 160..232 via formulas that reproduce the values from Rows 12..84 as in the previous section. As an optional feature (Figure 3.10), the fixed point A can be shown and the lines L generated using the multiple line procedure from Section 1.6. Alternate points come from the lower and upper curves.

```
       B     C       D      E      F      G      H     I     J     K      L     M    N     O
 1 CONCHOID
 2 Params:         Fix Pt,A:      Length:
 3 Stp:h0= 0.5  x0 =      0   k =        8
 4 Init:x= -18  y0 =     -3
 5 Curve Parameters:  y = a·x + b
 6      a =  0.3  b =      0
 :
10 Curve:                        Curve                                              Lo    Hi
11   n  θ:d    θ:r       r       x     y1     d    X:lo  Y:lo  X:hi  Y:hi           y2    y3
12   0                        -18.0  -5.4   18.2  -10.1  -4.3  -25.9  -6.5
13   1                        -17.5  -5.3   17.6   -9.6  -4.2  -25.4  -6.3
14   2                        -17.0  -5.1   17.1   -9.1  -4.1  -24.9  -6.1
 :
84  72                         18.0   5.4   19.9   10.8   2.0   25.2   8.8
85 Lower Curve:
86   0                        -10.1                                                -4.3
87   1                         -9.6                                                -4.2
 :
158 72                         10.8                                                 2.0
159 Upper Curve:
160  0                        -25.9                                                      -6.5
161  1                        -25.4                                                      -6.3
162  2                        -24.9                                                      -6.1
 :
232 72                         25.2                                                       8.8

B12: 0    B13: 1+B12                   F12: +D4    F13: +F12+D$3
G12: +D$6*F12+F$6                      H12: @SQRT((F12-F$3)^2+(G12-F$4)^2)
J12: +F12-H$3*(F12-F$3)/H12            K12: +G12-H$3*(G12-F$4)/H12
L12: +F12+H$3*(F12-F$3)/H12            M12: +G12+H$3*(G12-F$4)/H12
B86: +B12                              F86: +J12      N86: +K12
B160: +B12                             F160: +L12     O160: +M12
```

Figure 3.9. Conchoid (Base: Line).

Copy:		Graph:			
From	To	Series	Cells	Labels	Purpose
B13	B14..B84	X	F12..F232		x-axis
F13	F14..F84	1	G12..G84		base curve
G12..M12	G13..M84	2	N12..N158		lower curve
B86..N86	B87..N158	3	O12..O232		upper curve
B160..O160	B161..O232				

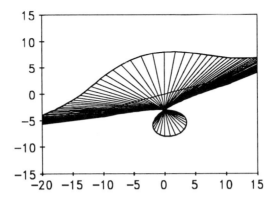

Figure 3.10. Conchoid with Lines.

Activities

1. Change with the location of the fixed point A, the line C, and the distance k.
2. Modify the model to show the construction lines and points of the definition.
3. Modify the base curve in the model to something other than a line, as for example, in Figures 3.11 and 3.12. Also try circles, ellipses, and other curves from Chapter 2 for the curve C.

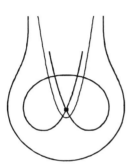

Figure 3.11. Conchoid of Parabola.

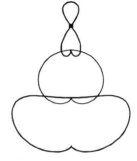

Figure 3.12. Conchoid of Cardioid.

3.3. CISSOIDS

Let C_1 and C_2 be curves and A a fixed point. A line L is drawn through the point A, intersecting the curves C_1 and C_2 at points Q_1 and Q_2, respectively, as in Figure 3.13. The locus of points P such that $AP = Q_1Q_2$ as Q_1 and Q_2 vary over the two curves is the *cissoid of C_1 and C_2 with respect to A*. Cissoids are discussed in Lockwood (1961), pp. 130 to 303; Teixeira (1971), pp. 1 to 26 of Tome I; Yates (1974), pp. 26 to 30; Lawrence (1972), pp. 98 to 100; and Pedoe (1976), pp. 246 to 248. The name cissoid means "ivy-shaped".

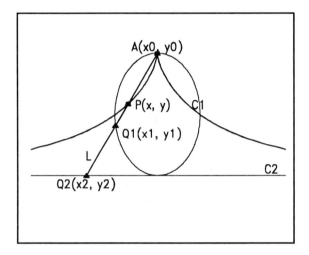

Figure 3.13. Cissoid (Base: Ellipse, Line).

The model of Figure 3.14 employs an ellipse $x = a\cos(\theta)$, $y = b\sin(\theta)$ for curve C_1, and a line $y = cx+d$ for curve C_2. The model's formulas assume that C_2 is always a line, while C_1 can be replaced by any curve. The values for a and b are entered in Cells D6 and F6, with c and d set in Cells D7 and F7, and $A = (x_0, y_0)$ in Cells G3..G4. Cell I3 calculates the value of a certain constant $k = cx_0+d-y_0$ that is used in several computations.

Cells B12..G84 generate C_1 as the base curve by a standard technique. Points $Q_1 = (x_1, y_1)$ on the ellipse C_1 are calculated in Columns F and G. For each point Q_1, the corresponding point $Q_2 = (x_2, y_2)$ must be located on the line C_2. Any point (x, y) on the line determined by the vector AQ_1 satisfies

$$(x,y) = (x_0, y_0) + t[x_1-x_0, y_1-y_0]$$

Since $Q_2 = (x_2, y_2)$ is such a point, there is a t_2 such that

$$x_2 = x_0 + t_2(x_1-x_0), \quad y_2 = y_0 + t_2(y_1-y_0)$$

Since this point lies on the line $y = cx+d$,

$$y_0 + t_2(y_1-y_0) = c[x_0 + t_2(x_1-x_0)]+d$$

After solving for t_2, the value of this variable for the first point is calculated in Cell I12 by

$$t_2 = [cx_0 + d - y_0]/[(y_1-y_0)-c(x_1-x_0)]$$

Next, let P = (x,y) be a point on the cissoid. From the definition, $[x-x_0,y-y_0] = [x_2-x_1,y_2-y_1]$, so

$$x = x_0 - x_1 + x_2 = x_0 - x_1 + [x_0 + t_2(x_1-x_0)]$$

Thus, the x-coordinate of the first point on the cissoid is found as

$$x = (2-t_2)x_0 + (t_2-1)x_1$$

in Cell J12, with the y-component computed similarly in Cell K12. The formulas in Row 12 are copied through Row 84.

The (x,y)-components for the cissoid's graph are reproduced from Rows 12..84 in Columns F and H of Rows 86..158 by entering +J12 and +K12 in Cells F86 and H86 and copying them through Row 158. The graph of the line C_2 is generated in Rows 160..162 from three points, while the fixed point is furnished in Row 163. A graph such as that of Figure 3.13 is produced by using the intervals $-6 \leq x \leq 6$, $-4 \leq y \leq 4$.

```
        B   C     D     E     F     G     H     I      J     K    L   M
 1 CISSOID (Ellipse/Line)
 2 Parameters:       Point:
 3 Step:θ0=     5    x0=         0  k =   -6.00
 4                   y0=         3
 5 Curves: C1: x = a·cos(θ), y = b·sin(θ), C2: y = cx + d
 6      a =     2    b=          3
 7      c =     0    d=         -3
 8 Constants:
 9 pi/180= 0.017
10 Curve:                       Elip  Ciss         Cissoid    L  Fix
11      n   θ:d  θ:rad    r      x    y1    y2  t2   X    Y  y3 y4
12      0    0   0.000   2.00  0.00               2.00  2.0  0.0
13      1    5   0.087   1.99  0.26               2.19  2.4 -0.3
14      2   10   0.175   1.97  0.52               2.42  2.8 -0.5
     :
84     72  360   6.283   2.00 -0.00               2.00  2.0  0.0
85
86      0                      2.00         0.00
87      1                      2.37        -0.26
88      2                      2.80        -0.52
     :
158    72                      2.00         0.00
159
160 Line:                -10                                    -3
161                       0                                    -3
162                       6                                    -3
163 Fixed Point:          0                                        3

I3: +D7*G3+F7-G4  D9: @PI/180     B12: 0     B13: 1+B12
C12: +D$3*B12                     D12: +D$9*C12
F12: +D$6*@COS(D12)               G12: +F$6*@SIN(D12)
I12: +I$3/(G12-G$4-D$7*(F12-G$3))
J12: (2-I12)*G$3+(I12-1)*F12   K12: (2-I12)*G$4+(I12-1)*G12
B86: +B12  F86: +J12               H86: +K12
F160: -10  L160: +D$7*F160+F$7  F161: 0    L161: +D$7*F161+F$7
F162: 6    L162: +D$7*F162+F$7  F163: +G3  M163: +G4
```

Figure 3.14. Cissoid (Base: Ellipse, Line).

94 Cissoids

Copy:		Graph:			
From	To	Series	Cells	Labels	Purpose
B13	B14..B84	X	F12..F163		x-axis
C12..K12	C13..K84	1	G12..G84		ellipse
B86..H86	B87..H158	2	H12..H158		cissoid
		3	L12..L162		line
		4	M12..M163		fixed point

Activities

1. Modify the model of Figure 3.14 to include construction lines and points.
2. Investigate the effects in changes in the parameters of the model. Notice that a given fixed point may generate two branches for the cissoid, as in Figure 3.15.

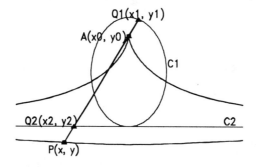

Figure 3.15. Cissoid (Base: Ellipse, Line).

3. Create cissoids using a variety of curves from Chapter 2 for C_1, as in Figures 3.16 and 3.17.

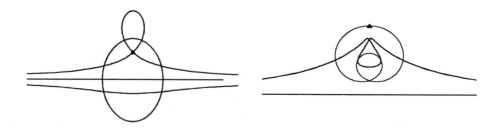

Figure 3.16. Cissoid (Base: Ellipse, Line). **Figure 3.17.** Cissoid (Base: Limaçon, Line).

3.4. PEDAL CURVES

Let C be a curve and A a fixed point, as in Figure 3.18. For each point Q on C a line L_1 is drawn tangent to C at Q. Also, a line segment L_2 is drawn from A and normal to L_1, intersecting L_1 at a point P, called the foot of the normal. As Q varies over the points of C, the locus of points P is the *pedal of C with respect to A*. Further ideas on pedals may be found in Lockwood (1961), pp. 152 to 155; Yates (1974), pp. 166 to 169; and Lawrence (1972), pp. 46 to 49.

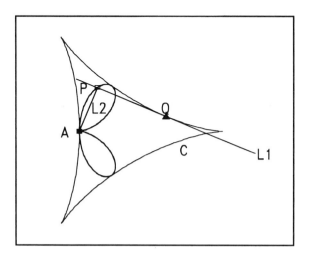

Figure 3.18. Pedal of Deltoid.

The model of Figure 3.19 uses the deltoid,

$$x = a[2\cos(\theta)+\cos(2\theta)], \quad y = a[2\sin(\theta)-\sin(2\theta)]$$

for the base curve C. The value of a is set in Cell D6, with the fixed point $A = (x_0, y_0)$ entered in Cells G3..G4. Using $a = 1$ and $A = (-1,0)$, the particular pedal produced is a bifolium.

In building models that use the reciprocals of of the slopes of lines, it is beneficial to avoid using values of θ that produce points whose tangent lines have slope 0, and whose normals therefore have an infinite slope. Thus, a small value is entered in Cell D7 that is used to produce a slight offset from the usual multiples of 5° generated for θ. Here an offset of 0.001° results in θ values of 0.001°, 5.001°, 10.001°,....

The curve C is formed in Columns F..G of Rows 12..84, as usual. Column H determines the slope of the tangent line at a point $Q = (x_1, y_1)$ on C by $dy/dx = (dy/d\theta)/(dx/d\theta)$. It is necessary to change the derivative formula if C is changed. Here the slope is given by

$$m = dy/dx = (\cos(2\theta)-\cos(\theta))/(\sin(2\theta)+\sin(\theta))$$

The equation of the tangent line at the point Q is

$$(1) \quad y-y_1 = m(x-x_1)$$

while that of the normal line passing through $A = (x_0, y_0)$, is

$$(2) \quad y - y_0 = (-1/m)(x - x_0)$$

The point $P = (X, Y)$ on the pedal curve where the lines intersect is derived by solving equations (1) and (2) simultaneously to give

$$X = (mx_1 - y_1 + y_0 + x_0/m)/(m + 1/m), \quad Y = y_1 + m(X - x_1)$$

The formulas that calculate these values are entered in Cells I12 and J12. When the formulas in Row 12 are copied through Row 84, the coordinates of the pedal appear in Columns I and J.

The (x, y) components of the pedal then are reproduced from Rows 12..84 into Columns F and K of Rows 86..158 by entering +I12 and +J12 in Cells F86 and K86 and copying them through Row 158. Row 159 is used to show the fixed point in a graph. The graph of Figure 3.18 uses Column F as the x-axis, Column G as the first y-series (the base curve), and Column K as the second y-series (the pedal).

```
        B         C        D    E    F       G         H        I       J      K     L
 1  PEDAL: Deltoid
 2  Parameters:              Fixed Point,A:
 3    Step: θ0=      5       x0 =    -1
 4                           y0 =     0
 5  Params: x=a[2cosθ+cos(2θ)],  y=a[2sinθ-sin(2θ)]
 6        a =        1
 7    offset =   0.001
 8  Constants:
 9    pi/180 =   0.017                                                         Fix
10  Curve:                         Delt  Slope                       Ped      Pt
11    n     θ:deg   θ:rad     x    y1     m          X      Y        y2       y3
12    0     0.001   0.000   3.00  0.00  -0.00     -1.00   0.00
13    1     5.001   0.087   2.98  0.00  -0.04     -0.99   0.17
 :
84   72   360.001   6.283   3.00  0.00  -0.00     -1.00   0.00
85  Pedal:
86    0                           -1.00                                      0.00
87    1                           -0.99                                      0.17
 :
158  72                           -1.00                                      0.00
159 Fixed Point:                  -1.00                                                0

B12: 0    B13: 1+B12      C12: +D7      C13: +C12+D$3
D12: +D$9*C12             F12: +D$6*(2*@COS(D12)+@COS(2*D12))
G12: +D$6*(2*@SIN(D12)-@SIN(2*D12))
H12: (@COS(2*D12)-@COS(D12))/(@SIN(2*D12)+@SIN(D12))
I12: (H12*F12-G12+G$4+G$3/H12)/(H12+1/H12)
J12: +G12+H12*(I12-F12)              D9:   @PI/180
B86: +B12   F86: +I12   K86: +J12    F159: +G3    L159: +G4
```

Figure 3.19. Pedal (Base: Deltoid).

Copy:		Graph:			
From	To	Series	Cells	Labels	Purpose
B13..C13	B14..C84	X	F12..F159		x-axis
D12..J12	D13..J84	1	G12..G84		deltoid
B86..K86	B87..K158	2	K12..K158		pedal curve
		3	L12..L159		foot of normal

Activities

1. Modify the model of Figure 3.19 to include construction lines and points.
2. Investigate the effect on the curves produced by varying the fixed point, A. Figures 3.20 to 3.23 show this with the astroid.

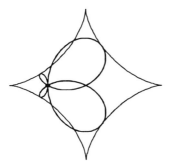

Figure 3.20. Pedal of Astroid.

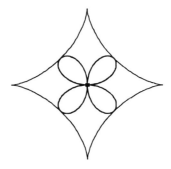

Figure 3.21. Pedal of Astroid.

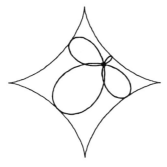

Figure 3.22. Pedal of Astroid.

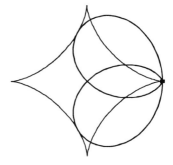

Figure 3.23. Pedal of Astroid.

3. Modify the model to create pedal curves using other base curves from Chapter 2, as in Figures 3.24 and 3.25. Remember to change the slope in Column H.

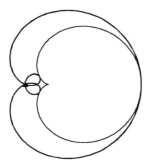

Figure 3.24. Pedal of Cardioid.

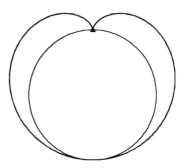

Figure 3.25. Pedal of Circle.

3.5. NEGATIVE PEDALS

Let C be a curve and A a fixed point. As shown in Figure 3.26, for each point Q of C, a line segment L_1 is drawn from A to Q. In addition, a line L_2 is drawn through Q and normal to L_1. As Q varies over the points of C, the envelope of the lines formed is the *negative pedal of C with respect to A*. The point A is called the foot of the negative pedal. Negative pedals are considered in Lockwood (1961), pp. 156 to 159, and Yates (1974), pp. 166 to 171.

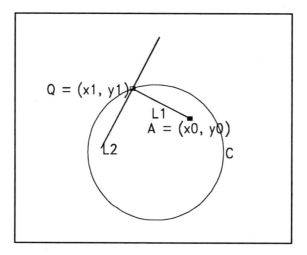

Figure 3.26. Negative Pedal (Base: Circle).

The model of Figure 3.27 uses the circle $r = a$ for the curve C as a special case of the curve $r = a + b\cos(\theta)$, with $a = 1$, $b = 0$. If A lies inside C, as in the example, the negative pedal is an ellipse (Figure 3.28). The values for a and b are entered in Cells D6 and F6, with the coordinates of the point $A = (x_0, y_0)$ in Cells G3..G4. Here, $A = (0.5, 0.5)$. The length k of the line segments that form the envelope is selected in Cell D4. As in Section 3.4, Cell D7 provides an offset for the values of θ.

The base curve C is created in Columns F..G of Rows 12..84 using the standard technique. The coordinates of the initial point of C, $Q = (x_1, y_1)$ are generated in Cells F12..G12. Cell H12 calculates the slope of the line segment QA as $m = (y_1 - y_0)/(x_1 - x_0)$. The slope of the perpendicular, $-1/m$, is used to calculate the angle α of the normal line in Cell I12 by @ATAN(-1/H12) as in Section 1.2. If the slope of the tangent line is infinite, then the normal line has slope 0. However, -1/H12 will return ERR. The formula in Cell I12 overcomes this difficulty by using

@IF(@ISERR(H12),0,@ATAN(-1/H12))

to return 0 when Cell H12 gives ERR.

Next, Columns J..M compute the coordinates of the two endpoints for the normal line segment whose length is k and direction is α, by

$$(x, y) = (x_1, y_1) \pm k[\cos(\alpha), \sin(\alpha)]$$

As in previous sections, the curve generated by the − sign in Columns J..K is called the lower curve, while the + sign in Columns L..M gives the upper curve. The formulas in Row 12 are copied through Row 84, and the block B12..M84 named as T. Finally, Rows 86..304 create the lines that form the envelope using the multiple line scheme of Section 1.6. The alternate pairs of points lie on the lower and upper curves.

```
         B      C       D      E     F      G      H    I    J     K    L    M    N
  1 NEGATIVE PEDAL: Circle/Cardioid
  2 Overall Parameters:   Foot:
  3   Step:θ0=        5     x0 =  0.5
  4   Len:  k=        1     y0 =  0.5
  5 Curve Parameters:  r = a + b·cos(θ)
  6       a =         1  b =    0
  7    Offset= 0.001
  8 Constants:
  9     pi/180=0.017
 10 Curve:                        Curve        Normal                        Lin
 11   n  θ(deg) θ(rad)   r     x    y1     m    α    xlo  ylo  xhi  yhi     y2
 12   0   0.001  0.00  1.00  1.00  0.00  -1.0  0.8  0.3  -0.7  1.7  0.7
 13   1   5.001  0.09  1.00  1.00  0.09  -0.8  0.9  0.4  -0.7  1.6  0.9
 14   2  10.001  0.17  1.00  0.98  0.17  -0.7  1.0  0.4  -0.7  1.5  1.0
  :
 84  72  360    6.28  1.00  1.00  0.00  -1.0  0.8  0.3  -0.7  1.7  0.7
 85 Negative Pedal:
 86   0     0            0.3                                            -0.7
 87   0     1            1.7                                             0.7
 88   0     2            NA                                              NA
 89   1     0            0.4                                            -0.7
 90   1     1            1.6                                             0.9
 91   1     2            NA                                              NA
  :
304  72     2            NA                                              NA

B12: 0   B13: 1+B12           C12: +D7   C13: +C12+D$3    D9: @PI/180
D12: +D$9*C12                 E12: +D$6+F$6*@COS(D12)
F12: +E12*@COS(D12)           G12: +E12*@SIN(D12)
H12: (G12-G$4)/(F12-G$3)      I12: @IF(@ISERR(H12),0,@ATAN(-1/H12))
J12: +F12-D$4*@COS(I12)       K12: +G12-D$4*@SIN(I12)
L12: +F12+D$4*@COS(I12)       M12: +G12+D$4*@SIN(I12)
B86: 0   B87: +B86+(C87=0)   C86: 0   C87: @MOD(1+C86,3)
F86: @CHOOSE(C86,@VLOOKUP(B86,$T,8),@VLOOKUP(B86,$T,10),@NA)
N86: @CHOOSE(C86,@VLOOKUP(B86,$T,9),@VLOOKUP(B86,$T,11),@NA)
```

Figure 3.27. Negative Pedal (Base: Circle).

Copy:		Graph:			
From	To	Series	Cells	Labels	Purpose
B13..C13	B14..C84	X	F12..F304		x-axis
D12..M12	D13..M84	1	G12..G84		circle
B87..C87	B88..C304	2	N12..M304		envelope lines
F86..N86	F87..N304				

Activities

1. Find the negative pedals that result from varying the foot or the parameters of the base curve, as shown in Figures 3.28 and 3.29. Show the foot in the graph.
2. Create envelopes for the negative pedals of other base curves, as in Figures 3.30 and 3.31. Use colors and line styles effectively.
3. Design a modification of the model that condenses the layout.

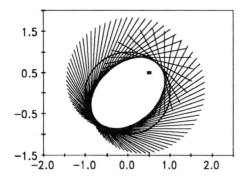

Figure 3.28. Negative Pedal of Circle. **Figure 3.29.** Negative Pedal of Cardioid.

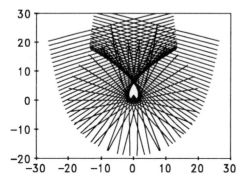 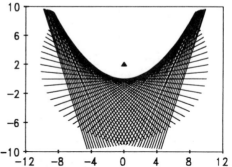

Figure 3.30. Negative Pedal of Parabola. **Figure 3.31.** Negative Pedal of Line.

3.6. INVERSE CURVES

Let C be a curve, A a fixed point, and r a constant distance. For each point Q of C a line L is drawn from A through Q as in Figure 3.32. A point P on L such that $AQ \cdot AP = r^2$ is called the *inverse of Q with respect to A*. The locus of all such points P as Q varies over C is called the *inverse of C with respect to A*. P satisfies the proportion $AQ:r = r:AP$. Each point that lies on the circle C_0 of radius r whose center is A is its own inverse. C_0 is called the *circle of inversion*. The inverses of points in the interior of C_0 lie in the exterior of C_0, and conversely. Inversion is considered in Lockwood (1961), pp. 177 to 181; Yates (1974), pp. 127 to 134; Lawrence (1972), pp. 43 to 46; and Courant and Robbins (1941), pp. 140 to 145. This topic is also addressed using complex variables in Chapter 5.

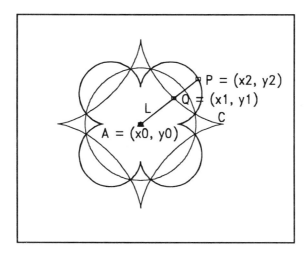

Figure 3.32. Inverse of Astroid.

The model of Figure 3.33 uses an astroid (see Section 2.6) for the curve C. Its equation is $x = a\cos^3(\theta)$, $y = a\sin^3(\theta)$. The value of the astroid parameter a is entered in Cell D6, with $A = (x_0, y_0)$ in Cells F3..F4, and the radius r of the circle of inversion in Cell H3. The value of r^2 is determined in Cell H4 as +H3*H3. An incremental step size of $\theta_0 = 2.5°$ (Cell D3) is used in drawing C in order to produce a smoother curve.

The base curve C is formed in Columns F..G of Rows 12..156. The coordinates of points $Q = (x_1, y_1)$ of the astroid are generated in Columns F..G. In Cell H12, the square of the distance d from A to the first point Q of C is calculated as

$$d^2 = (x_1 - x_0)^2 + (y_1 - y_0)^2$$

To complete the block, the formulas in Row 12 are copied down through Row 156.

The inverse curve is created in Rows 158..302. To construct the point $P = (x_2, y_2)$ corresponding to a point Q, the line segment from $A = (x_0, y_0)$ to $Q = (x_1, y_1)$ must be extended. A unit vector in this direction is

$$(1/d)[x_1 - x_0, y_1 - y_0]$$

Because $AP \cdot AQ = r^2$ and $AQ = d$, the distance from A to P is r^2/d. Thus, P is determined by

$$(x_2, y_2) = (x_0, y_0) + (r^2/d^2)[x_1 - x_0, y_1 - y_0]$$

and Cells F158 and I158 compute

$$x_2 = x_0 + (r^2/d^2)(x_1 - x_0), \quad y_2 = y_0 + (r^2/d^2)(y_1 - y_0)$$

These formulas are copied through Row 302.

The model also creates the circle of inversion, which is constructed in the usual way using polar coordinates in Rows 304..376, with the x- and y-values produced in Columns F and J, respectively. Row 377 is used to include the center point A in the graph.

102 *Inverse Curves*

```
          B      C      D       E      F       G      H       I      J      K
 1 INVERSE: Astroid
 2 Parameters:           Inversion Circle:
 3     Step:θ0=  2.5  x0=   0      R =    2
 4                      y0=   0      R^2=   4
 5 Curve Parameters: x = a[cos(θ)]^3, y = a[sin(θ)]^3
 6         a =    3
 7
 8 Constants:
 9     pi/180=0.017
10 Curve:                             Astr            Inv  I/Cir  Ctr
11   n  θ(deg) θ(rad)   r      x      y1      d²      y2    y4    y3
12   0    0.0  0.000          3.00   0.00    9.00
13   1    2.5  0.044          2.99   0.00    8.95
14   2    5.0  0.087          2.97   0.00    8.80
  :
156 144  360.0  6.283         3.00   0.00    9.00
157 Inverse:
158   0                       1.33                    0.00
159   1                       1.34                    0.00
160   2                       1.35                    0.00
  :
302 144                       1.33                   -0.00
303 Circle of Inversion:
304   0     0   0.00          2.00                          0.00
305   1     5   0.09          1.99                          0.17
306   2    10   0.17          1.97                          0.35
  :
376  72   360   6.28          2.00                         -0.00
377 Center:                      0                                        0

H4: +H3*H3      C12: +D$3*B12     B12: 0    B13: 1+B12
D9: @PI/180  D12: +D$9*C12     F12: +D$6*@COS(D12)^3
G12: +D$6*@SIN(D12)^3          H12: (F12-F$3)^2+(G12-F$4)^2
B158: +B12                     F158: +F$3+H$4*(F12-F$3)/H12
I158: +F$4+H$4*(G12-F$4)/H12
B304: +B12    C304: 5*B304     D304: +D$9*C304
F304: +F$3+H$3*@COS(D304)      J304: +F$4+H$3*@SIN(D304)
F377: +F3                      K377: +F4
```

Figure 3.33. Inverse Curve (Base: Astroid).

Copy:		Graph:			
From	To	Series	Cells	Labels	Purpose
B13	B14..B156	X	F12..F377		x-axis
C12..H12	C13..H156	1	G12..G156		astroid
B158..I158	B159..I302	2	I12..I302		inverse curve
B304..J404	B305..J376	3	K12..K377		fixed point
		4	J12..J376		inversion circle

<u>Activities</u>

1. Investigate the effects of a change in location of the point A.
2. Modify the model to sketch inverses of other curves from Chapter 2, as shown in Figures 3.34 and 3.35. Astroids and lemniscates, in particular, produce interesting curves.

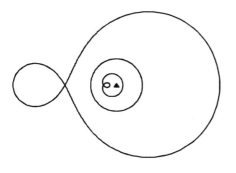
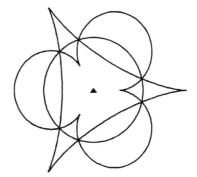

Figure 3.34. Inverse of Limaçon.   **Figure 3.35.** Inverse of Deltoid.

3. Illustrate the fact that the inverse of a line is a circle, unless the line passes through the center of the circle of inversion, in which case the inverse is a line.
4. Modify the model of Figure 3.33 to include the construction lines and points.

3.7. PARALLEL CURVES

Let C be a curve and k a constant. For each point Q on C a line L_1 is drawn tangent to C at Q, and a ray L_2 is drawn from Q that is normal to L_1 as in Figure 3.36. A point P that is k units from Q is located on L_2. The locus of such points P as Q varies over C is a curve that is *parallel* to C. The topic of parallel curves is found in Yates (1974), pp. 157 to 159, and Bruce et al. (1990), pp. 120 to 129.

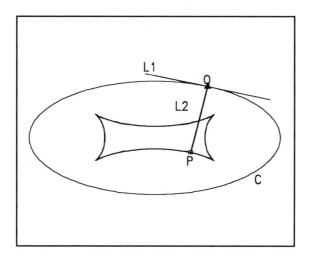

Figure 3.36. Parallel Curve (Base: Ellipse).

A given curve C determines two parallel curves, one located on each "side" of C. The model of Figure 3.37 is designed to select one of these curves, depending upon the sign of k. The model uses an ellipse, $x = a\cos(\theta)$, $y = b\sin(\theta)$ for the base curve C. Values for a and b are entered in Cells D6 and F6, with k set in Cell G3.

104 *Parallel Curves*

The base curve C is created in Columns F..G of Rows 12..84, with the incremental step size, θ_0, for θ set in Cell D3. The coordinates of the initial point, $Q = (x_1,y_1)$, of C are computed in Cells F12 and G12. The slope, m, of the tangent line at Q (which depends upon the curve C) is found in Cell I12 by the derivative, $dy/dx = (dy/d\theta)/(dx/d\theta)$, or

$$m = b\cos(\theta)/(-a\sin(\theta)) = -b/(a\tan(\theta))$$

The slope of the corresponding normal line is $-1/m$. As discussed in Section 3.5, the direction angle α of the normal is computed in Cell J12 as

$$@IF(@ISERR(I12),0,@ATAN(-1/I12))$$

There are two points on each perpendicular line that are k units from Q,

$$(x_2,y_2) = (x_1,y_1) \pm k[\cos(\alpha),\sin(\alpha)]$$

If, say, the negative sign were always selected for y, then the points generated would always lie below the base curve. This is not exactly what is wanted. Instead, the more complex IF..THEN..ELSE construction shown is used to select the correct point. The formulas for the first point (X,Y) of the parallel curve are entered in Cells K12 and L12 as shown, and then copied through Row 84.

Rows 86..158 reproduce the (x,y) coordinates of the points of the parallel curve. The formulas +K12 and +L12 are entered in Cells F86 and H86 and copied down the columns. Columns F and H define the graph of the parallel curve.

```
          B      C       D      E     F       G      H      I       J       K     L
 1 PARALLEL CURVES: Ellipse/Circle
 2 Overall Parameters Distance:
 3    Step:θ0=       5       k =  -2.4
 4
 5 Curve Parameters:  x = a·cos(θ),   y = b·sin(θ)
 6        a =     4   b=   2
 7
 8 Constants:
 9    pi/180=0.017
10 Curve:                        Curve   Par          Perp
11    n   θ:deg  θ:rad     x      y1     y2     m     Angle    X     Y
12    0      0   0.000   4.00    0.00                 ERR    0.00   1.6   0.0
13    1      5   0.087   3.98    0.17                 -5.7   0.17   1.6  -0.2
14    2     10   0.175   3.94    0.35                 -2.8   0.34   1.7  -0.5
   :
84   72    360   6.283   4.00   -0.00                 ****  -0.00   1.6  -0.0
85 Parallel Curve:
86    0                   1.60           0.00
87    1                   1.62          -0.24
88    2                   1.68          -0.45
   :
158  72                   1.60          -0.00

D9:  @PI/180       B12: 0           B13: 1+B12
C12: +D$3*B12      D12: +D$9*C12    F12: +D$6*@COS(D12)
G12: +F$6*@SIN(D12)                 I12: -F$6/(D$6*@TAN(D12))
J12: @IF(@ISERR(I12),0,@ATAN(-1/I12))
K12: +F12+G$3*@ABS(@COS(J12))*@IF(G13<G12,-1,1)
L12: +G12+G$3*@ABS(@SIN(J12))*@IF(F13<F12,1,-1)
B86: +B12          F86: +K12        H86: +L12
```

Figure 3.37. Parallel Curve (Base: Ellipse).

Copy:		Graph:			
From	To	Series	Cells	Labels	Purpose
B13	B14..B84	X	F12..F158		x-axis
C12..L12	C13..L84	1	G12..G84		ellipse
B86..H86	B87..H158	2	H12..H158		parallel curve

Activities

1. By changing k, investigate how the nature of the parallel curves can change dramatically, as illustrated in Figures 3.38 to 3.41. Observe how the sign of k determines the curve that is produced.

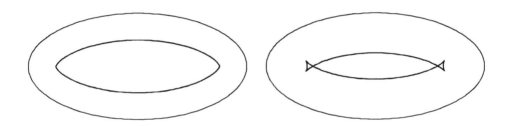

Figure 3.38. Parallel to Ellipse, k = -1.0.   **Figure 3.39.** Parallel to Ellipse, k = -1.5.

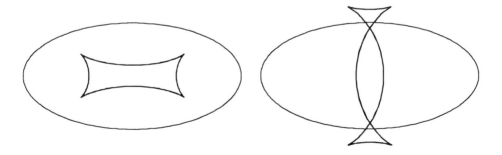

Figure 3.40. Parallel to Ellipse, k = -2.4.   **Figure 3.41.** Parallel to Ellipse, k = -4.5.

2. Create models that use other curves such as lemniscates, hyperbolas, astroids, and deltoids as the base curve. See Figures 3.42 through 3.47 and Yates (1974), p. 157.
3. Modify the model to show simultaneously both of the two possible parallel curves that lie on either side of the curve.

Figure 3.42. Parallel to Parabola.

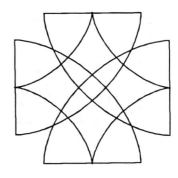

Figure 3.43. Parallel to Astroid.

Figure 3.44. Parallel to Rose.

Figure 3.45. Parallel to Rose.

Figure 3.46. Parallel to Rose.

Figure 3.47. Parallel to Rose.

3.8. EVOLUTES

Let C be a curve. From Figure 3.48, for each point Q of C a line L_1 is drawn tangent to C at Q, and a second line L_2 is drawn through Q and normal to L_1. As Q varies over the points of C, the envelope of the resulting family of lines is called the *evolute of C*. The evolute is defined optionally as the locus of the centers of curvature of the points Q of C. Discussions of evolutes are found in Lockwood (1961), pp. 166 to 168; Taylor (1992), pp. 236 to 240; Yates (1974), pp. 86 to 92; and Bruce et al. (1990), pp. 129 to 139.

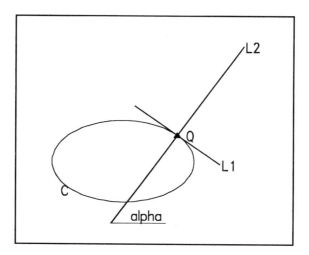

Figure 3.48. Evolute (Base: Ellipse).

The model of Figure 3.49 uses an ellipse, $x = a\cos(\theta)$, $y = b\sin(\theta)$ for the curve C. The values for a and b are entered in Cells D6..D7. The length k of the line segments that generate the envelope is set in Cell D4. The particular envelope that is generated in Figure 3.50 is called a *Lamé curve*.

```
          B    C      D   E   F      G       H     I      J      K     L     M     N
       1 EVOLUTE: Ellipse
       2 Overall Parameters:
       3 Step:θ=    5
       4 Len: k=    8
       5 Curve Parameters:  x = a·cos(θ),  y = b·sin(θ)
       6     a =    5
       7     b =    3
       8 Constants:
       9 π/180=0.017
      10 Curve:                      Eli       Nor                              Line
      11  n  θ:d   θ:r   r    x   y1    m     α   X-lo  Y-lo  X-hi  Y-hi   y2
      12  0   0   0.00  5.0  0.0   ERR  0.0  -3.0   0.0  13.0   0.0
      13  1   5   0.09  5.0  0.3  -6.9  0.1  -2.9  -0.9  12.9   1.4
      14  2  10   0.17  4.9  0.5  -3.4  0.3  -2.8  -1.7  12.6   2.8
          :
      84 72 360   6.28  5.0  0.0  ****  0.0  -3.0   0.0  13.0  -0.0
      85
      86 Envelope Lines:
      87  0   0           NA                                            NA
      88  0   1         -3.0                                           0.0
      89  0   2         13.0                                           0.0
      90  1   0           NA                                            NA
      91  1   1         -2.9                                          -0.9
      92  1   2         12.9                                           1.4
          :
     305 72   2         13.0                                          -0.0

     D9:  @PI/180                       B12: 0   B13: 1+B12
     C12: +D$3*B12    D12: +D$9*C12    F12: +D$6*@COS(D12)
     G12: +D$7*@SIN(D12)               H12: -D$7/(D$6*@TAN(D12))
     I12: @IF(@ISERR(H12),0,@ATAN(-1/H12))
     J12: +F12-D$4*@COS(I12)           K12: +G12-D$4*@SIN(I12)
     L12: +F12+D$4*@COS(I12)           M12: +G12+D$4*@SIN(I12)
     B87: 0   B88: +B87+(C88=0)        C87: 0   C88: @MOD(1+C87,3)
     F87: @IF(C87=0,@NA,@VLOOKUP(B87,$T,6+2*C87))
     N87: @IF(C87=0,@NA,@VLOOKUP(B87,$T,7+2*C87))
```

Figure 3.49. Evolute (Base: Ellipse).

108 Evolutes

Copy:		Graph:			
From	To	Series	Cells	Labels	Purpose
B13	B14..B84	X	F12..F305		x-axis
C12..M12	C13..M84	1	G12..G84		ellipse
B88..C88	B89..C305	2	N12..N305		envelope lines
F87..N87	F88..N305				

The base curve C is formed in Columns F and G of Rows 12..84 as in the previous sections. Cell H12 determines the slope of the tangent at the curve's initial point $Q = (x_1, y_1)$ of C as $m = dy/dx = (dy/d\theta)/(dx/d\theta)$. The specific formula depends upon the particular curve C. For the ellipse,

$$m = b\cos(\theta)/(-a\sin(\theta)) = -b/(a\tan(\theta))$$

The creation of the rest of the model follows from Section 3.5. The angle of the normal line, $\alpha = \arctan(-1/m)$, for the first point of C is calculated in Cell I12. Upper and lower endpoints of line segments that extend k units in either normal direction from Q are constructed in Columns J..M by

$$(x_1, y_1) \pm k[\cos(\alpha), \sin(\alpha)]$$

The formulas in Row 12 are copied down through Row 84, and the block B12..M84 named T. The normal lines for the envelope are formed in Rows 87..305, with points coming from the lower and upper (x, y) values in the block T.

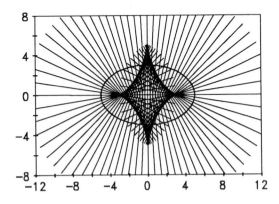

Figure 3.50. Evolute of Ellipse.

Activities

1. Modify the model to create evolutes for other base curves as in Figures 3.51 and 3.52. The appropriate derivative formula must be entered in Column H. The evolute of the cardioid in Figure 3.52 is another cardioid. (See Bruce et al., 1990, pp. 132 to 134.)

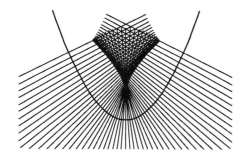

Figure 3.51. Evolute of Parabola.

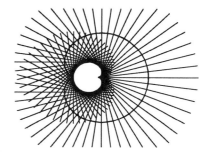

Figure 3.52. Evolute of Cardioid.

2. Extend the model of Figure 3.49 to create the curves formed by connecting the endpoints of the normal line segments. Investigate the effects of changes in k.
3. The related concept of the involute is harder to implement. See Lockwood (1961), p. 169, for a method of constructing an approximation of the involute of a curve.

3.9 - CAUSTICS

Let C be a curve and A a fixed point. As in Figure 3.53, for each point Q of C the line segment L_1 is drawn from A to Q. Next, a line segment L_2 is constructed, with its center at Q and lying on the line produced by reflecting L_1 in the tangent to C at Q. As Q varies over the points of C, the lines L_2 form the envelope of the *caustic of C with respect to A*. (See Lockwood, 1961, pp. 182 to 185; Yates, 1974, pp. 15 to 20; Bruce et al., 1990, pp. 196 to 206.)

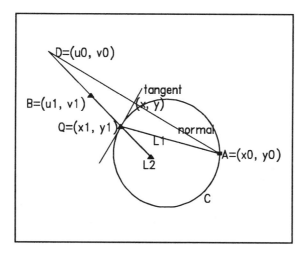

Figure 3.53. Caustic (Base: Circle).

110 Caustics

The caustic of Figure 3.54 is created from the model of Figure 3.55 using a circle of radius a for C as a special case of the ellipse $x = a\cos(\theta)$, $y = b\sin(\theta)$, with $A = (5,0)$ together with $a = b = 5$. The values for a and b are entered in Cells D6..D7, with the point $A = (x_0,y_0)$ set in Cells G3..G4. The length k of the segment L_2 is set in Cell K4. As in Section 3.4, a small angle offset for the θ values is entered in Cell D4 to produce values of θ at which the slopes of tangent lines are neither 0 nor infinite.

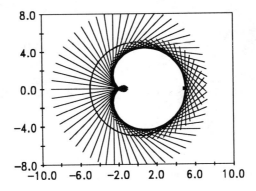

Figure 3.54. Caustic of Circle.

```
          B     C      D     E    F      G      H    I    J    K    L    M    N    O    P
 1 CAUSTIC CURVE  (Ellipse/Circle)
 2 Parameters:          Fixed Pt(A):
 3    Step:θ=     5     x0=   5
 4    Offset=   0.01    y0=   0         len:k=  4
 5 Curve Parameters: x = a·cos(θ),  y = b·sin(θ)
 6       a =     5
 7       b =     5
 8 Constants:
 9    pi/180= 0.017
                            Curve       Reflect  Dist                    Line
10 Curve:
11    n  θ:deg   θ:r    x     y1    m    u0   v0    d  xlo  ylo  xhi  yhi  y2
12    0   0.01  0.00  5.00  0.00  ***  5.0  0.0  0.0  5.0  4.0  5.0 -4.0
13    1   5.01  0.09  4.98  0.44  -11  5.0  0.0  0.4  4.5  4.4  5.5 -3.5
14    2  10.01  0.17  4.92  0.87   -6  5.1  0.0  0.9  3.9  4.7  6.0 -3.0
   :
84   72  360    6.28  5.00  0.00  ***  5.0  0.0  0.0  5.0  4.0  5.0 -4.0
85 Lines:
86    0    0          5.00                                              4.0
87    0    1          5.00                                             -4.0
88    0    2          NA                                                NA
89    1    0          4.46                                              4.4
90    1    1          5.50                                             -3.5
91    1    2          NA                                                NA
   :
303  72    1          5.00                                             -4.0

B12: 0           B13: 1+B12           C12: +D4        C13: +C12+D$3
D9:  @PI/180     D12: +C12*D$9        F12: +D$6*@COS(D12)
G12: +D$7*@SIN(D12)                   H12: -D$7/(D$6*@TAN(D12))
I12: 2*(H12^2*F12+G$3+H12*(G$4-G12))/(1+H12^2) -G$3
J12: 2*(H12^2*G$4+G12+H12*(G$3-F12))/(1+H12^2) -G$4
K12: @SQRT((I12-F12)^2+(J12-G12)^2)
L12: +F12-(K$4/K12)*(I12-F12)    M12: +G12-(K$4/K12)*(J12-G12)
N12: +F12+(K$4/K12)*(I12-F12)    O12: +G12+(K$4/K12)*(J12-G12)
B86: 0   B87: +B86+(C87=0)       C86: 0   C87: @MOD(1+C86,3)
F86: @CHOOSE(C86,@VLOOKUP(B86,$T,10),@VLOOKUP(B86,$T,12),@NA)
P86: @CHOOSE(C86,@VLOOKUP(B86,$T,11),@VLOOKUP(B86,$T,13),@NA)
```

Figure 3.55. Caustic (Base: Circle).

Copy:		Graph:			
From	To	Series	Cells	Labels	Purpose
B13..C13	B14..C84	X	F12..F303		x-axis
D12..O12	D13..O84	1	G12..G84		circle
B87..C87	B88..C303	2	P12..P303		envelope lines
F86..P86	F87..P303				

As usual, the base curve C is created in Columns F and G of Rows 12..84, with the initial point $Q = (x_1, y_1)$ of C in Cells F12..G12. The slope of the tangent at Q is calculated in Cell H12 as

$$m = dy/dx = b\cos(\theta)/(-a\sin(\theta)) = -b/(a\tan(\theta))$$

Again, the formula for the derivative must be changed to use the model with another curve.

The point (x,y) at which the normal from A intersects the tangent satisfies the equations

$$y - y_1 = m(x - x_1), \quad y - y_0 = (-1/m)(x - x_0)$$

Solving these simultaneously gives

$$x = [m(y_0 - y_1) + m^2 x_1 + x_0]/[1 + m^2], \quad y = [m^2 y_0 + y_1 + m(x_0 - x_1)]/[1 + m^2]$$

Thus, the point $D = (u_0, v_0)$ that is the reflection of the point A in the tangent line can be determined as

$$(u_0, v_0) = (x, y) + [x - x_0, y - y_0], \text{ or } u_0 = 2x - x_0, \; v_0 = 2y - y_0$$

These expressions are combined with the values found for x and y to create the formulas in Cells I12..J12.

The distance d from Q to D is obtained in Cell K12 as

$$d = \sqrt{(u_0 - x_1)^2 + (v_0 - y_1)^2}$$

Thus, a vector from Q of length k in the correct direction is

$$(k/d)[u_0 - x_1, v_0 - y_1]$$

Consequently, the coordinates of the inner and outer reflections can be determined as

$$(x_1, y_1) \pm (k/d)[u_0 - x_1, v_0 - y_1].$$

Cells L12..M12 use the minus sign to obtain a point on the inner curve as

$$x = x_1 - (k/d)(u_0 - x_1), \quad y = y_1 - (k/d)(v_0 - y_1)$$

while points on the outer curve are determined similarly in Cells N12..O12 by using the

112 *Caustics*

plus sign. The formulas of Row 12 are copied through Row 84, with the block B12..O84 named as T.

Finally, Rows 86..303 use the standard multiple line procedure from Section 1.6 to generate the reflected rays. Alternate points come from the lower (Columns L and M) and upper (Columns N and O) curves. The model need be extended only slightly to include the fixed-point in the graph.

Activities

1. Experiment with the effect of the location of A on the caustics that are produced. A few graphs are shown with a circle of radius 5 in Figures 3.56 to 3.59. Some of the caustics have two components. It is helpful to adjust k to produce clear pictures. In Figure 3.57, the point A is placed at "infinity" by setting x to a large number, say, 1.0E10.
2. Modify the model to include construction lines and points.
3. Create additional interesting curves defined by the endpoints of the rays. You may wish to eliminate the rays from the graph as in Figure 3.60.
4. Generate caustics for other curves from Chapter 2, such as parabolas and ellipses. A cardioid is used in Figure 3.61.

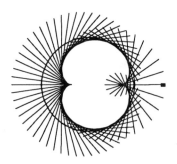

Figure 3.56. Caustic: P = (8,0).

Figure 3.57. Caustic, P = (∞,0).

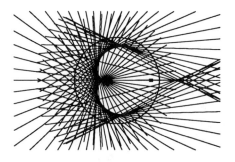

Figure 3.58. Caustic, P = (4,0).

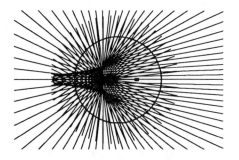

Figure 3.59. Caustic, P = (2.2,0).

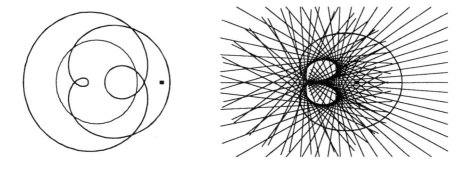

Figure 3.60. Caustic Embellished.   **Figure 3.61.** Caustic of Cardioid.

3.10. AN APPROXIMATION MODEL

Many geometric constructions require creating families of either tangent or normal lines at each of the points of a given base curve, C. Previous examples have implemented such processes using derivatives. However, there is an straightforward method for producing good approximations of the tangents without needing to use derivatives. In this method, at each point P_n of the curve, the slope of the chord that is determined by the two immediate surrounding points, P_{n-1} and P_{n+1}, is used to approximate the derivative at P_n. This slope is then used to approximate the tangent line at P_n by drawing a line segment that is centered at P_n, parallel to the chord, and extends in k units in each direction from P_n, as shown in Figure 3.62.

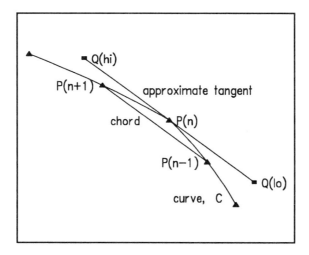

Figure 3.62. Tangent Approximation.

The model of Figure 3.65 employs a circle of radius 1 as the base curve with segments of length $k = 1$ to produce the graph of Figure 3.63. This model also can be used to produce normal lines as well as tangents. Figure 3.64 was produced by using normals to an ellipse, $x = 0.6\cos(\theta)$, $y = 1.2\sin(\theta)$. In fact, the line segments can be set to any angle α.

To construct the model, the length k of the line segments is entered in Cell E4. Cell G4 provides another parameter, the degree measure of the angle α of the generated line relative to the corresponding chord. Using $\alpha = 0°$ produces tangents, while $\alpha = 90°$ gives normals. The radian measure of α is calculated in Cell H4 by @RADIANS(G4).

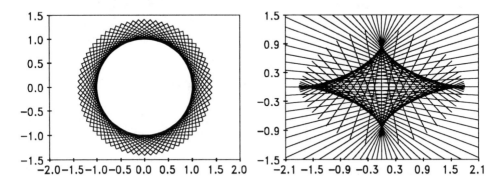

Figure 3.63. Approximation (Tangent).   **Figure 3.64.** Approximation (Normals).

The model is formed in two segments. Rows 12..85 create the base curve C, here, a circle of radius 1, as in previous examples. Column B serves as a point counter. The indicated formulas are entered in Cells C12..G12 and copied down their columns through Row 85. Including the extra Row 85, as shown, permits a line to be drawn at $\theta = 0$. The formula for θ entered in Cell C12 includes a small offset of 0.0001. Including this offset avoids the formation of pairs of points that produce an infinite slope. Columns F and G of this block create a circle through the relationship $(x,y) = (\cos(\theta),\sin(\theta))$. The model can be modified to generate curves from other parametric equations as well.

The angle of the initial chord, for $n = 1$, is calculated in Cell H13 as

$$\beta = \arctan[(y_2-y_0)/(x_2-x_0)] + \alpha$$

where α is the added angle of Cell H4. This formula is then copied down Column H to produce the corresponding angles at successive points. Next, the coordinates of one end, Q_{lo}, of the approximation line are determined in Cells I13..J13 by finding the point that is $-k$ units in direction β from P_1, or $(x_1-k\cos(\beta),y_1-k\sin(\beta))$. The coordinates of the other endpoint, Q_{hi}, are found similarly in Cells K13..L13 as $(x_1+k\cos(\beta),y_1+k\sin(\beta))$. These formulas are copied down through Row 85, and the block B12..L85 named as T.

Next, the family of lines determined by the curve points is formed in Rows 90..305 using the standard multiple line technique of Section 1.6. The indicated formulas are entered in Rows 90..91 and copied appropriately. The alternate points of each line are the coordinates of the "lo" and "hi" endpoints of the line segments, as read from Columns I..J and K..L of block T through table lookup functions.

Once the model is created, it is easy to modify it to implement another curve and to create normals rather than tangents.

```
      B     C      D      E       F       G       H       I      J      K      L
   1 TANGENT APPROXIMATION:
   2 Parameters:
   3    Line length:    Angle:   deg     rad
   4          k =     1      α =     0       0
   :
  11    n   θ:d    θ:r            r       x       y       ß      xlo    ylo    xhi    yhi
  12    0    0    0.00                 1.00    0.00
  13    1    5    0.09                 1.00    0.09   -1.48    0.91   1.08   1.08  -0.91
  14    2   10    0.17                 0.98    0.17   -1.40    0.81   1.16   1.16  -0.81
  :
  83   71  355    6.20                 1.00   -0.09    1.48    0.91  -1.08   1.08   0.91
  84   72  360    6.28                 1.00   -0.00   -1.57    1.00   1.00   1.00  -1.00
  85   73  365    6.37                 1.00    0.09      NA      NA     NA     NA     NA
  86                                     NA      NA      NA      NA     NA     NA     NA
  87
  88 Lines:
  89         n    mod                    x       y
  90         0     0                    NA      NA
  91         0     1                  0.00    0.00
  92         0     2                  0.00    0.00
  93         1     0                    NA      NA
  94         1     1                  0.91    1.08
  95         1     2                  1.08   -0.91
  :
 303        71     0                    NA      NA
 304        71     1                  0.91   -1.08
 305        71     2                  1.08    0.91

H4:  @RADIANS(G4)                B12: 0    B13: 1+B12
C12: 5*B12+0.0001                D12: @RADIANS(C12)
F12: @COS(D12)                   G12: @SIN(D12)
H13: @ATAN((G14-G12)/(F14-F12))+H$4              F86..H86: @NA
I13: +F13-E$4*@COS(H13)          J13: +G13-E$4*@SIN(H13)
K13: +F13+E$4*@COS(H13)          L13: +G13+E$4*@SIN(H13)
C90: 0   C91: +C90+(D91=0)   D90: 0   D91: @MOD(1+D90,3)
F90: @CHOOSE(D90,@NA,@VLOOKUP(C90,$T,7),@VLOOKUP(C90,$T,9))
G90: @CHOOSE(D90,@NA,@VLOOKUP(C90,$T,8),@VLOOKUP(C90,$T,10))
```

Figure 3.65. Tangent Approximation (Base: Circle).

Copy:		Graph:			
From	To	Series	Cells	Labels	Purpose
B13	B14..B85	X	F90..305		x-axis
C12..G12	C13..G85	1	G90..305		lines
H13..L13	H14..L85				
C91..D91	C92..D305				
F90..G90	F91..G305				

The ideas of this section have an intriguing application. The March 18, 1940, issue of *Life Magazine* (pp. 43 to 44) contains an article discussing some ideas of Professor George Birkhoff of Harvard University on drawing pictures by using only straight lines. The drawing of the fox of Figure 3.67 was suggested by one of the pictures drawn there using traditional means by a student, David Middleton. The spreadsheet drawing here has been created by expanding both segments of this section's model. In the upper part of the model, the base sketch of the fox shown in Figure 3.66 is created by implementing several curves in the table T. For example, the sides of the head are translations of the curves $x = \pm y^4$, the top of the head is an arc of a circle, the ears are rotated parabolas, and the nose and face are portions of lemniscates. Points of these curves are generated in T and numbered consecutively. Next, the insert and copy commands are used to generate the multiple line section forming Figure 3.67 as a result. The ensuing model will be quite large.

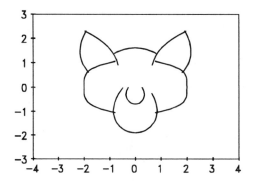

Figure 3.66. Fox Base Curves.

Figure 3.67. Fox via Lines.

Activities

1. Create the model of this section. Then modify it for use with various curves to construct approximations of families of tangents and normals. Also, vary α to examine other families that are produced. Include parameters in the modified models.
2. Create families of tangents to lines using the derivative (Taylor, 1992, pp. 218 to 223).
3. Design an approximation technique for drawing a tractrix, a curve and construction that is discussed in Lockwood (1961), pp. 118 to 124.
4. Use the ideas of the model to create drawings via lines as discussed above.

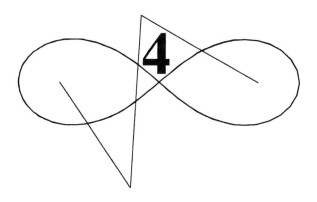

Chapter 4

ANIMATION

The models in the previous chapters are designed so that it is easy to modify their parameters and investigate the changes that are produced. The models in this chapter use circular references to extend this process a step further to generate an animation effect. The numerical and graphical output of these models is updated automatically by repeatedly pressing the recalculation key. Various results are produced by this process: one curve may be deformed continually by systematic changes in a parameter, another curve is traced out in a point-by-point fashion, some figures expand and contract, circles roll while other figures slide, and a rotating ray sweeps out a drawing. The techniques demonstrated can be quite valuable for illustrating the step-by-step construction of a curve or families of curves.

The examples shown are especially effective when implemented on spreadsheets that allow both numerical and graphical output to be displayed simultaneously through embedded graphs. In certain spreadsheets, especially *Excel* 4.0, the animation effect is particularly smooth and dramatic. *Quattro Pro* works well too, especially with version 2.0, which manifests only a quick "blink" when the curve is redrawn. The process also works with *Lotus* 2.3, 3.0 and other versions. However, even in the other spreadsheets these models provide a very productive way for updating curves and constructions. Such examples can be used to produce effective displays for both classroom and professional presentations through the use of a liquid crystal display (LCD) projection device.

4.1. TRACING A CURVE

It can be quite instructive to trace out a curve in an animated, point-by-point manner. Such a technique is especially effective for examining curves that loop over themselves. The model in this section can be produced from only slight modifications in the basic parametric model of Section 2.2. Here, the *four-leafed rose*, $r = \cos(2\theta)$, is traced out, as depicted in Figures 4.1 and 4.2.

118 Tracing a Curve

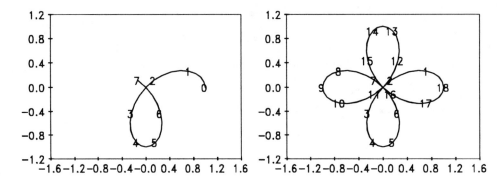

Figure 4.1. Four-Leafed Rose, $r = \cos(2\theta)$. **Figure 4.2.** Four-Leafed Rose, $r = \cos(2\theta)$.

First, the entire construction of the more general curve $r = a + b\cos(d\theta)$ is carried out in Cells B12..G84 just as it was in Figure 2.5. This model uses a more abbreviated function, but the same effect is achieved easily by using the function and parameters of Figure 2.5 and setting $c = e = 0$. The values of the curve's parameters are set in Rows 6..7, with the formula for r, +D$6+F$6*@COS(F$7*D12), entered in Cell E12 and copied through Row 84. However, instead of simply sketching the entire curve by using Columns F and G for the x- and y-series, the model uses circular references to regenerate successively points (X,Y) of the curve in Columns I and J. The number, k, of new points that are displayed with each recalculation is set in Cell D4. Here, $k = 4$. As usual, Column B provides a count, n, for the points on the curve.

```
           B      C       D       E       F       G       H       I       J
 1  PARAMETRIC EQUATIONS: Trace
 2  Parameters:      Recalculate:
 3  Step:θ0=     5  Start>       1
 4  Plot:k =     4  N =          4
 5  Curve Parameters:  r = a + b·cos(d·θ)
 6          a =      0   b =     1
 7                       d =     2
 8  Constants:
 9  pi/180= 0.017
10  Curve:                              Curve          Trace
11   n   θ:deg  θ:rad    r       x       y  Label    X       Y
12   0     0   0.000   1.00    1.00    0.00     0  1.00    0.00
13   1     5   0.087   0.98    0.98    0.09        0.98    0.09
14   2    10   0.175   0.94    0.93    0.16        0.93    0.16
15   3    15   0.262   0.87    0.84    0.22        0.84    0.22
16   4    20   0.349   0.77    0.72    0.26     1  0.72    0.26
17   5    25   0.436   0.64    0.58    0.27        NA      NA
  :
84  72   360   6.283   1.00    1.00   -0.00    18  NA      NA

F4:   @IF(F3=0,0,D4+F4)            D9:   @PI/180
B12:  0        B13:  1+B12         C12:  +D$3*B12
D12:  +D$9*C12                     E12:  +D$6+F$6*@COS(F$7*D12)
F12:  +E12*@COS(D12)               G12:  +E12*@SIN(D12)
H12:  @IF(@MOD(B12,D$4),"",B12/D$4)
I12:  @IF(F$4<B12,@NA,F12)  J12:  @IF(F$4<B12,@NA,G12)
```

Figure 4.3. Curve Tracing.

Copy:		Graph:			
From	To	Series	Cells	Labels	Purpose
B13	B14..B84	X	I12..I84		x-axis
C12..J12	C13..J84	1	J12..J84	H12..H84	curve

Cell F3 is used to initialize the process. The total number, N, of points that are plotted is calculated in Cell F4 by @IF(F3=0,0,D4+F4). When Cell F3 is set to 0, N becomes 0, and only the first point is computed. When F3 is set to any nonzero value (say, 1), and each time that the recalculation key is pressed thereafter, N (Cell F4) is increased by k (Cell D4). In addition, with each recalculation k more points are generated for the graph in Columns I..J. This permits a viewer to see the curve "grow". An animation effect is produced if the graph is displayed on screen simultaneously with the data.

To achieve this updating effect, the x- and y-values in Columns I,J are generated by entering the formulas

I12: @IF(F$4<B12,@NA,F12), J12: @IF(F$4<B12,@NA,G12)

and copying them through Row 84. Each of these formulas returns @NA until the value of the counter N (Cell F$4) reaches or exceeds the value of n in Column B. At that time the (x,y) values are reproduced in Columns I..J from the corresponding entries in Columns E..F of those rows. Finally, Column H provides interior labels for the graph that count every k-th point.

Activities

1. Create similar models for other curves of interest. The curves from Chapter 2 provide good examples. Also, condense the models to require fewer columns by combining computations. For example, compute $\cos(n\theta_0\pi/180)$ in a single cell. Vary the size of k to discover an appropriate pace for the movement.
2. Modify the model's recalculation scheme so that points are traced one at a time. Then use the spreadsheet's recalculation command to set the number of iterations that are performed each time the recalculation key is pressed.

4.2. VARYING PARAMETERS

This section describes how to implement another animated format in which the graph of a curve is continuously readjusted, or deformed, as one parameter is changed systematically. The model uses a circular reference to update the parameter a of the curve $r = a + b\cos(d\theta)$ with each recalculation. This process is illustrated in Figures 4.4 to 4.7, starting with a cardioid.

The model is described in Figure 4.8. The initial value, a_0, for a and the step increment, h, for changes in the parameter a are entered in Cells H3 and H4. The main body of the model of Figure 4.8 is created as it was for Figure 4.3. Cell F3 is used to initialize the

120 *Varying Parameters*

model, with the value of the parameter *a* computed in Cell D6 by @IF(F3=0,H3,D6+H4). When Cell F3 is set to 0, the value of *a* in Cell D6 is a_0. When Cell F3 is set to a nonzero value, and with each ensuing recalculation, the value of *a* in Cell D6 is increased by *h*. The effects of changes in *a* can be followed through the graph, as shown below.

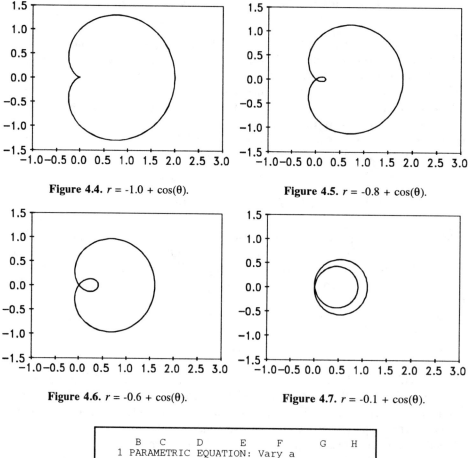

Figure 4.4. $r = -1.0 + \cos(\theta)$.

Figure 4.5. $r = -0.8 + \cos(\theta)$.

Figure 4.6. $r = -0.6 + \cos(\theta)$.

Figure 4.7. $r = -0.1 + \cos(\theta)$.

Figure 4.8. Cardioid, Parameter *a* Varies.

Copy:		Graph:			
From	To	Series	Cells	Labels	Purpose
B13	B14..B84	X	F12..F84		x-axis
C12..G12	C13..G84	1	G12..G84		curve

Activities

1. Modify the model of Figure 4.8 by selecting another parameter, for example, b or d, to be altered continually. Create similar models for other curves from Chapter 2.
2. Create a model with a graph that contains several y-series, so that the curves for the past few values of the parameter a are preserved and displayed with each recalculation, as older curves are replaced. Employ different colors or line styles for the graphs.
3. Animate the tangent line model of Figure 1.14 so that the selected point starts at the left side and moves to the right along the curve.
4. The graph of a conic section can be given in polar form by $r = ed/(1+e\cos(\theta))$, where e is the *eccentricity* of the curve. Create an animated model that examines the changing nature of such a curve as e gradually decreases from $e = 4$ to $e = 0$.
5. Modify the model of this section by letting x vary in step increments and produce graphs of functions $y = f(x)$ that include parameters that are continually updated to produce animation effects. For example,

   (a) Provide motion in the trigonometric functions such as $\cos(ax)$, $\cos(a+x)$, $a\cos(x)$
   (b) Continuously deform polynomials such as $(x-a)(x-2a)(x-3a)$, $(x-a)(x-a+1)(x-a+2)$
   (c) Examine the changing nature of logistic functions such as $k/(1+(k-1)e^{-kax})$
   (d) Vary the normal curve, $y = e^{-(x/a)^2/2}/(ab)$, where a is standard deviation, $b^2 = 2\pi$

6. Create an animated model to illustrate the concept of resonance that occurs in the study of second-order differential equations. For example, show the changing nature of the function $y = a\cos(\omega t) + b\cos(\omega_0)/(\omega^2-\omega_0^2)$ as ω approaches ω_0. Typical output is shown in Figures 4.9 and 4.10.

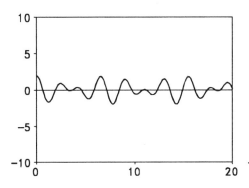
Figure 4.9. $a = 1$, $b = 4$, $\omega \approx 3$, $\omega_0 = 2$.

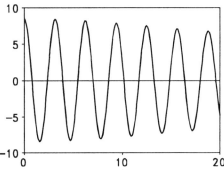
Figure 4.10. $a = 1$, $b = 4$, $\omega \approx 2.12$, $\omega_0 = 2$.

4.3. CONIC SECTIONS VIA CIRCLES

In Section 2.10 graphs of the conic sections were created from parametric equations. Here, animated models produce these curves from their definitions. The properties of these curves that are used in the next two sections are discussed in most of the references on curves. (See Finney and Thomas, 1990, pp. 646 to 660; Lockwood, 1961, pp. 2 to 33; Yates, 1974, pp. 36 to 55.)

4.3.A. Ellipse

An *ellipse* is the set of all points P for which the sum of the distances from two fixed points, called *foci*, to P is a constant, S. This model uses circular references to create a pair of circles with centers at the foci $(-c,0)$, $(c,0)$. As the spreadsheet is recalculated, the radius of the left circle continually increases, while the radius of the right circle decreases, so that the sum of their radii remains S. As the circles change, their points of intersection trace an ellipse. Figures 4.11 through 4.14 use $c = 2$, $S = 6$. The graphical output can be enhanced by the effective use of color.

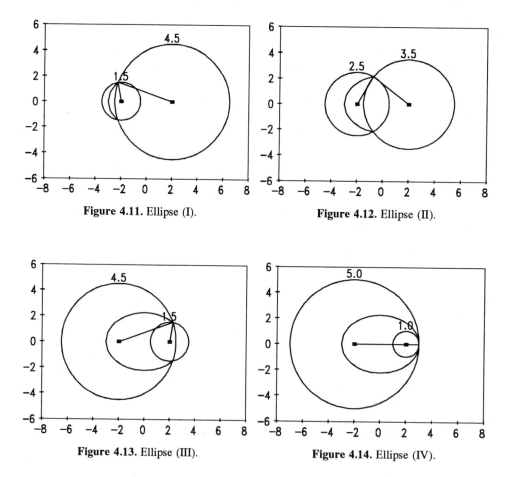

Figure 4.11. Ellipse (I).   Figure 4.12. Ellipse (II).

Figure 4.13. Ellipse (III).   Figure 4.14. Ellipse (IV).

A model that produces these curves is shown in Figure 4.15. Values for c and S are set in Cells D4 and D5. In this example the sum of the radii is chosen as $S = 6$. In Cell D3 the step size increment d_0 for successive changes in the radii is assigned. Here, $d_0 = 0.1$. Cell G4 counts iterations i by @IF(G3=0,0,1+G4). In Cell G5 the number N/i of radii increments to be carried out with each recalculation is entered, and the cumulative number N of increments in the radii is computed in Cell G6 as +G4*G5. The output shown indicates that there have been four recalculations of five points each for a total of 20 points plotted.

Cell G3 is used to initialize the process. When 0 is entered in Cell G3, i is set to 0, and no points are created. When Cell G3 is changed to a nonzero value, and with each subsequent recalculation, i is incremented by 1, and N is increased by an additional N/i.

```
        B       C       D       E       F       G       H       I       J       K
 1  ELLIPSE via Circles
 2  Overall Parameters:
 3    Step:d0=  0.1             START>    1
 4    Foci:c=   2                 i =     4   count
 5    Sum: S=   6                N/i=     5   changes/count
 6                                 N =   20   changes
 :
 9    pi/180=  0.017            Left    Right           Up      Low
10  Left Circle:                Circ    Circ   Foc     Ell      Ell   Radii
11    n   θ:deg θ:rad     x      y1      y2     y3      y4       y5      y6
12    0    0    0.00   0.00    0.00
13    1   15    0.26  -0.07    0.52
14    2   30    0.52  -0.27    1.00
 :
35   23  345    6.02  -0.07   -0.52
36   24  360    6.28   0.00   -0.00
37  Right Circle:
38    0    0    0.00   6.00            0.00
39    1   15    0.26   5.86            1.04
40    2   30    0.52   5.46            2.00
 :
62   24  360    6.28   6.00           -0.00
63  centers and radii:
64                    -2.00            0                         0
65                    -1.50                                           1.94
66                     2.00            0                         0
67  ellipse:
68    0               -4.50                           ERR      ERR
69    1               -4.35                           ERR      ERR
 :
77    9               -3.15                           ERR      ERR
78   10               -3.00                          0.00     0.00
79   11               -2.85                          0.70    -0.70
 :
88   20               -1.50                          1.94    -1.94
89   21                 NA                            NA       NA
 :
120  52                 NA                            NA       NA

G4: @IF(G3=0,0,1+G4)          G6: +G4*G5         D9:  @PI/180
B12: 0      B13: 1+B12         C12: 15*B12       D12: +C12*D$9
E12: -D$4+G$6*D$3*@COS(D12)    F12: +G$6*D$3*@SIN(D12)
B38: +B12   C38: +C12          D38: +D12
E38: +D$4+(D$5-D$3*G$6)*@COS(D38)
G38: (D$5-D$3*G$6)*@SIN(D38)
E64: -D4    H64: 0     K64: 0
E65: +D5*(2*G6*D3-D5)/(4*D4)
K65: @SQRT((G6*D3)^2-(E65+D4)^2)
E66: +D4    H66: 0     K66: 0     B68: 0   B69: 1+B68
E68: @IF(B68>G$6,@NA,D$5*(2*B68*D$3-D$5)/(4*D$4))
I68: @IF(B68>G$6,@NA,@SQRT((B68*D$3)^2-(E68+D$4)^2))
J68: -I68
```

Figure 4.15. Ellipse via Circles.

Copy:		Graph:			
From	To	Series	Cells	Labels	Purpose
B13	B14..B36	X	E12..E120		x-axis
C12..F12	C13..F36	1	F12..F36		left circle
B38..G38	B39..G62	2	G12..G62		right circle
B69	B70..B120	3	H12..H66		foci
E68..J69	E69..J120	4	I12..I120		upper ellipse
		5	J12..J120		lower ellipse
		6	K12..K66		radii

The (x,y)-coordinates of the left circle are created using steps of $15°$ in Columns E..F of Rows 12..36 via the standard procedure. Smoother curves will result by using smaller steps and by increasing the size of the model. This circle is created with a variable radius of Nd_0 and center $(-c,0)$. The indicated formulas calculate points on the circle as

$$x = -c + Nd_0\cos(\theta), \; y = Nd_0\sin(\theta)$$

These formulas are entered in Row 12 and copied through Row 36. Both N and the radius of the left circle increase with each recalculation. The right circle, with radius $S-Nd_0$ and center $(c,0)$, is created similarly in Columns E and G of Rows 38..62 using the formulas

$$x = c + (S-Nd_0)\cos(\theta), \; y = (S-Nd_0)\sin(\theta)$$

With each calculation, its radius $S-Nd_0$ decreases while the sum of the two radii remains S.

Rows 68..120 successively generate points on the ellipse that is traced by the points of intersection of the two circles. Column B produces a count n for the points. When Cell G3 is set to 0, all of the (x,y)-values except the first in Columns E and I are @NA. When Cell G3 is set to a nonzero value, (x,y)-values are displayed for those points for which $n \leq N$. The points of intersection are found by simultaneously solving the equations

$$(x+c)^2 + y^2 = (nd_0)^2, \; (x-c)^2 + y^2 = (S-nd_0)^2$$

to find that

$$x = S(2nd_0-S)/(4c), \; y = \pm\sqrt{(nd_0)^2-(x+c)^2}$$

and selecting the + sign for y to generate points on the upper half of the ellipse. Because N is incremented by steps of size N/i (Cell G5), N/i new points are displayed with each recalculation. The last point displayed for each recalculation is for $n = N$.

For each x, the two circles generally intersect in two points. The y-coordinates of the lower half of the ellipse are determined in Column J as the negative of the entry in Column I. When the circles fail to intersect, ERR is generated in Columns I and J. Rows 64..66 reproduce the foci, determine the current point of intersection of the circles in Row 65, and draw the radii.

4.3.B. Hyperbola

A *hyperbola* is the set of all points P such that the difference in the distances from two fixed points, called *foci*, to P is a constant. Graphs produced from a model similar to that of Figure 4.15 to generate one branch of a hyperbola are shown in Figures 4.16 and 4.17. Here, the foci are at (-2,0), (2,0) and the difference is 1.5. Both of the circles grow with each recalculation. The construction of the model for the hyperbola is left as an exercise.

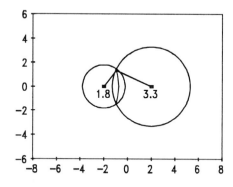
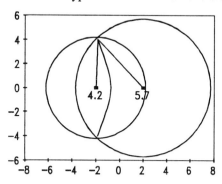

Figure 4.16. Hyperbola (I). **Figure 4.17.** Hyperbola (II).

4.3.C. Parabola

A *parabola* is the set of all points P that are equidistant from a fixed point, called the *focus*, and a line L, called the *directrix*. In the example of Figure 4.18 the focus is (-c,0) and L is $x = c$. In this particular implementation shown, $c = 1$. The operation of this model is similar to that for the ellipse, generating the graphs of Figure 4.19 and 4.20. The graph is updated one point at a time.

However, this model uses yet another animation technique. The values of the incremental size d_0 and the focus c are set in Cells D3 and D4. Cell G3 is used to initialize the model, while Cell G4 counts iterations N. The circle with center at $(-c,0)$ and variable radius Nd_0 is created in Rows 12..36, while the line $x = c - Nd_0$ is formed in Rows 38..39. With each recalculation, the radius of the circle increases by d_0 units and the line moves to the left by d_0 units. The points of intersection of the successive circles and lines determine a parabola.

Rows 41..47 create the foci, the upper of the two current points of intersection, and the connecting radii. For the current value of N the (x,y)-coordinates of the point of intersection of the line and the circle are found in Row 46 by solving the simultaneous equations

$$(x+c)^2 + y^2 = (Nd_0)^2, \quad x = c - Nd_0$$

to obtain

$$x = c - Nd_0, \quad y = \sqrt{(Nd_0)^2 - (x+c)^2}$$

Rows 49..101 generate the parabola. Column B provides a count n for the points. The first point is calculated in Row 49. Initially, @NA is displayed for the other points (x,y). After Cell G3 is set to a nonzero value, when the iteration counter N is equal to n the (x,y) coordinates of the point found in Row 46 are reproduced in Columns E and I of that n-th row. Otherwise, the entries do not change. The y-coordinates for the lower points of intersection are calculated in Column J as the negatives of the entries in Column I. With

Conic Sections via Circles

each recalculation another point of the parabola is generated. As before, ERR appears when the line and the circle fail to intersect.

```
       B     C     D      E        F      G     H    I     J     K
 2  Parameters:                           PARABOLA via Circle/Line
 3  Step:d0=   0.2         Start>    1
 4  Focus:c=   1           N  =      7  iterations
 :
 8  Constants:
 9      pi/180=0.017                    Dist      Top   Low
10  Curve:                       Circle Line Dir  Par   Par  Line
11     n   deg   rad      x       y1    y2   y3   y4    y5    y6
12     0    0   0.00    0.40    0.00
13     1   15   0.26    0.35    0.36
14     2   30   0.52    0.21    0.70
 :
36    24  360   6.28    0.40   -0.00
37  line:
38                     -0.40          -6.00
39                     -0.40           6.00
40
41  focus:             -1.00           0.00
42                       NA             NA
43                      1.00          -6.00
44                      1.00           6.00
45  radii:             -1.00                                  0.00
46       intersect>    -0.40                                  1.26
47                      1.00                                  1.26
48  parabola:
49     0                1.00                 ERR   ERR
 :
53     4                0.20                 ERR   ERR
54     5               -0.00                0.00  0.00
55     6               -0.20                0.89 -0.89
56     7               -0.40                1.26 -1.26
57     8                 NA                  NA    NA
 :
101   52                 NA                  NA    NA

G4: @IF(G3=0,0,1+G4)               B12: 0         B13: 1+B12
C12: 15*B12                        D12: +C12*D$9
E12: -D$4+D$3*G$4*@COS(D12)        F12: +D$3*G$4*@SIN(D12)
E38: +D4-G4*D3    G38:  -6         E39: +E38      G39: -G38
E41: -D4          H41:   0         E43: +D4       H43:  -6
E44: +E43         H44:   6         E45: +E41      K45: +H41
E46: +D4-D3*G4    K46: @SQRT((G4*D3)^2-(E46+D4)^2)
E47: -E45         K47: +K46        B49: 0         B50: 1+B49
E49: +D4-B49*D3  E50: @IF(G$3=0,@NA,@IF(B50=G$4,E$46,E50))
I49: @SQRT((B49*D$3)^2-(E49+D$4)^2)            J49: -I49
I50: @IF(G$3=0,@NA,@IF(B50=G$4,K$46,I50))
```

Figure 4.18. Parabola via Circles.

Copy:		Graph:			
From	To	Series	Cells	Labels	Purpose
B13	B14..B36	X	E12..E104		x-axis
C12..F12	C13..F36	1	F12..F36		left circle
B50..I50	B51..I101	2	G12..G39		parallel line
J49	J50..J101	3	H12..H44		focus, directrix
		4	I12..I101		upper parabola
		5	J12..J101		lower parabola
		6	K12..K47		distance lines

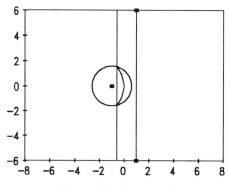
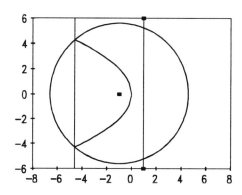

Figure 4.19. Parabola (I). **Figure 4.20.** Parabola (II).

Activities

1. Modify each of the models so that the circles are plotted in increments of 5°, and thereby produce smoother curves.
2. Create an animated model for a hyperbola. Generate both branches.
3. Use Column K of Figure 4.18 to create interior labels to show the distances from the current point on the parabola to the focus and the directrix.
4. Each of the curves in this section have foci on the x-axis. Modify the models so that the foci are on the y-axis.
5. Modify selected models from Chapter 3 to employ an animated format. One additional step of the construction process is carried out with each recalculation. Include the current locations of the construction lines and points.

4.4. TRACING CONIC SECTIONS

This section provides a second animated format for tracing the conic sections. This technique, again, comes directly from definitions of the conic sections.

4.4.A. Ellipse

An *ellipse* is the set of all points such that the sum of the distances from two fixed points, called *foci*, is a constant. Here, the constant sum of the distances is denoted by $2a$, and the foci are placed at $(-c,0)$ and $(c,0)$. This model describes the classical pencil and string construction of the ellipse. If the ends of a piece of string of length $2a$ are attached at the foci and a pencil is used to pull the string taut, then as the pencil is moved it traces the path of the ellipse, as illustrated in Figures 4.21 to 4.24.

128 *Tracing Conic Sections*

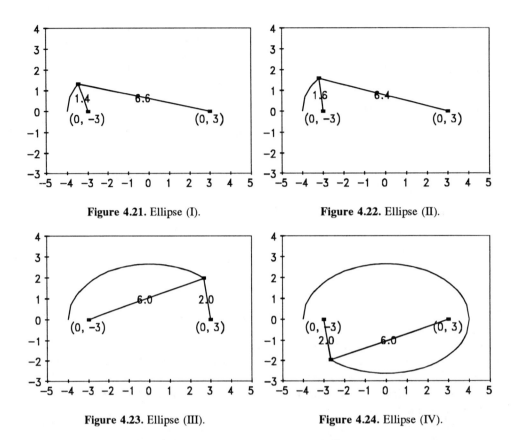

Figure 4.21. Ellipse (I).

Figure 4.22. Ellipse (II).

Figure 4.23. Ellipse (III).

Figure 4.24. Ellipse (IV).

In the model of Figure 4.26, the values of a and c are entered in Cells D6 and D7. The number, k, of points to plot with each recalculation is selected in Cell D4. This number controls the rate at which the curve is traced out. Cell G3 initializes the model. Cell G4 counts the cumulative number N of points generated. After Cell G3 is set to a nonzero value, more of the ellipse is swept out with each recalculation, as shown above with $c = 3$ and $2a = 8$.

The ellipse is created in Cells B12..F132. As usual, Column B counts points on the curve. From Figure 4.25, it follows that with $S = 2a$ the x-axis vertices are located at $(0, \pm a)$.

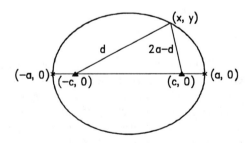

Figure 4.25. Ellipse Notation.

Column D creates a series of values that represent the distance d_n from the n-th point (x,y) on the ellipse to $(-c,0)$. These values for d_n start in Cell D12 at $a-c$ (the distance to the left vertex) and increase in steps of h_0 to $a+c$ (the distance to right vertex) in Cell D72, after which they decrease in steps of h_0 back to $a-c$ in Cell D132. The value for the increment h_0 (here, 0.1) is entered in Cell D3.

```
       B         C     D      E     F      G     H      I     J
 1 ELLIPSE
 2 Parameters:
 3   Step:h0=    0.1          Start>  1
 4   Plot:k=     2            N =     2
 5 Curve Parameters:
 6   Dist:a=     4            N(p)=   60
 7   Foci:c=     3
 :
10 Curve:                 Total Curve  Move Curve  Lines Foci
11     n    sign  d(n)    x     y      x     y     y2    y3
12     0     1    1.0   -4.00  0.00  -4.00  0.00
13     1     1    1.1   -3.87  0.68  -3.87  0.68
14     2     1    1.2   -3.73  0.95  -3.73  0.95
15     3     1    1.3   -3.60  1.15   NA    NA
 :
71    59     1    6.9    3.87  0.68   NA    NA
72    60    -1    7.0    4.00 -0.00   NA    NA
73    61    -1    6.9    3.87 -0.68   NA    NA
 :
132  120    -1    1.0   -4.00  0.00   NA    NA
133 Additional Features:
134                                   3.0         0.0   0.0
135                                  -0.4         0.5
136                                  -3.7         0.9   0.9
137                                  -3.4         0.5
138                                  -3.0         0.0   0.0

G4:   @IF(G3=0,0,D4+G4)            G6:  2*D7/D3
D9:   @PI/180                      B12: 0    B13: 1+B12
C12:  @IF(B12<G$6,1,-1)            D12: +D6-D7
E12:  +D$6*(D12-D$6)/D$7           D13: +D12+C12*D$3
F12:  +C12*@SQRT(D12^2-(E12+D$7)^2)
G12:  @IF(B12<=G$4,E12,@NA)        H12: @IF(B12<=G$4,F12,@NA)
G134: +D7                          I134: 0   J134: 0
G135: (G134+G136)/2                I135: (I134+I136)/2
G136: @VLOOKUP(G4,$T,3)            I136: @VLOOKUP(G4,$T,4)
J136: +I136   G137: (G136+G138)/2  I137: (I136+I138)/2
G138: -G134   I138: 0              J138: +I138
```

Figure 4.26. Ellipse via String.

Copy:		Graph:			
From	To	Series	Cells	Labels	Purpose
B13	B14..B132	X	G12..G138		x-axis
C12	C13..C132	1	H12..H132		ellipse
D13	D14..D132	2	I12..I138	K12..K137	lines
E12..H12	E13..H132	3	J12..J138	L12..L138	foci/point

Cell G6 computes the number of points N_p to be plotted in each of the upper and lower halves of the ellipse as

$$N_p = [(a+c)-(a-c)]/h_0 = 2c/h_0$$

Points on the ellipse occur in pairs: (x,y) and $(x,-y)$. Column C generates the sign for y. The expression @IF(B12<G$6,1,-1) is entered into Cell C12 and copied. Thus, 1 is created when $n < N_p$, and -1 when $n \geq N_p$. In Column D the distances d_n are formed by entering $a-c$ (+D6-D7) into Cell D12, and subsequently computing the $(n+1)$st value from the n-th value by $d_{n+1} = d_n \pm h_0$, where the sign comes from Column C.

The (x,y)-coordinates of the ellipse are calculated in Columns E..F. If d and e denote the distances from (x,y) to $(-c,0)$ and $(c,0)$, respectively, and their sum is $d+e = 2a$, then

$$(x+c)^2 + y^2 = d^2, \quad (x-c)^2 + y^2 = e^2$$

$$x^2 + 2xc + c^2 + y^2 = d^2, \quad x^2 - 2xc + c^2 + y^2 = e^2$$

$$4xc = d^2 - e^2 = d^2 - (2a-d)^2 = 4ad - 4a^2$$

so that $x = (d-a)a/c$. In addition, $y^2 = d^2 - (x+c)^2$, so that

$$y = \pm\sqrt{d^2-(x+c)^2}$$

The indicated formulas are entered in Cells E12..F12 and copied through Row 132, with the block B12..F132 named as T.

The animated curve is constructed in Columns G,H using a standard technique:

G12: @IF(B12<=G$4,E12,@NA)   H12: @IF(B12<=G$4,F12,@NA)

The graph is created using Column G for the x-axis series and Column H for the first y-series. Except for Row 12, the entries for x and y in these columns initially show NA. After Cell G3 is set to a nonzero value, with each recalculation the value of N (Cell G4) is increased by k so that k more points are calculated and graphed. Rows 134..138 provide lines to depict the distances from the foci to the current point on the curve by using Column I as the second y-series. Intermediate points are used to display the distances. The foci and the moving point are generated as symbols in the third y-series (Column J). The distances and the foci can be provided through interior labels with the second and third y-series as

K135: 2*D6-K137,   K137: @VLOOKUP(G4,T,2)
L134: "(0,"&@STRING(D7,0)&")",   L138: "(0,-"&@STRING(D7,0)&")".

4.4.B. Hyperbola and Parabolas

Hyperbolas and parabolas can be constructed similarly to produce the graphs of Figures 4.27 through 4.30.

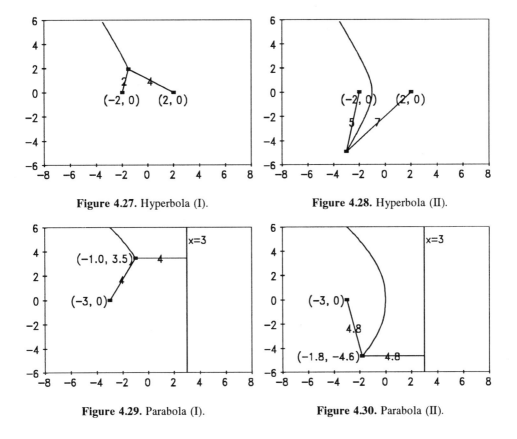

Figure 4.27. Hyperbola (I).

Figure 4.28. Hyperbola (II).

Figure 4.29. Parabola (I).

Figure 4.30. Parabola (II).

<u>Activities</u>

1. Using the ellipse model, experiment with various values of c for a given value of a. Notice that as c approaches 0, the ellipse nearly becomes a circle. As c varies, it may be necessary to change h_0 and use the insert and copy commands to generate more points to produce satisfactorily smooth curves.

2. In computing y at a vertex, the formula in Column F should calculate the square root of 0. However, at this point computer rounding may generate a slightly negative value rather than 0 for which the square root is to be found. This will cause the @SQRT function to return ERR, and in turn produce a gap in the graph. One way to overcome this difficulty is to change the entry in Cell F12 in Figure 4.26 slightly to

    F12: +C12*@SQRT(D12^2-(E12+D$7)^2+1.0E-12)

    and then copy it down Column F.

3. Design and implement the corresponding models for hyperbolas and parabolas.
4. Use the insert and copy commands to extend the various graphs, and to generate more points to produce smoother curves.
5. Two points start at the top of a circle of radius a and center at the origin. The points move at constant, though different, rates. One point moves down the y-axis, while the other point proceeds clockwise along the circumference so that both points reach the x-axis at the same time. The *quadratrix* is the curve that is formed as the locus

of the points of intersection of the radius that is drawn to the point on the circumference and the horizontal line that is drawn from the point on the y-axis. When the point on the y-axis has moved t units, the angle that the radius makes with the positive x-axis is $\theta = \pi(1-t)/2$, and the corresponding point on the quadratrix has coordinates $y = 1-t$, $x = y/\tan(\theta)$. Create an animated model for the construction, as in Figure 4.31.

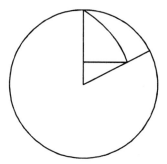

Figure 4.31. Quadratrix Construction.

4.5 - CYCLOIDS AS ROLLING CIRCLES

A *cycloid* is the path traced by a point P on the circumference of circle that rolls along a straight line without slipping. This section presents an animated model to generate a cycloid via such a rolling circle. It also generates *curtate* and *prolate cycloids* by tracing the paths of points inside and outside of the rolling circle. These curves are discussed in Courant and Robbins (1941), pp. 152 to 154; Bruce et al. (1990), pp. 112 to 114; Lockwood (1961), pp. 80 to 89; Yates (1974), pp. 65 to 70; Teixeira (1971), pp. 133 to 154 of Tome II; and in most calculus books.

4.5.A. Cycloids

This model traces the path of the point P originally on the bottom of the circle C of radius 1 and center (0,1). Illustrative output is shown in Figures 4.32 to 4.35. The model of Figure 4.36 is initialized by setting Cell G3 to 0. After Cell G3 is set to a nonzero value, and with each recalculation thereafter, C rolls further along the positive x-axis.

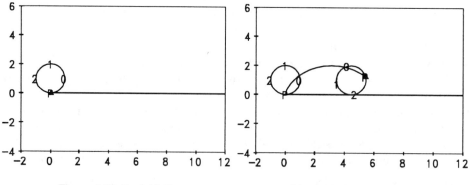

Figure 4.32. Cycloid (I).   **Figure 4.33.** Cycloid (II).

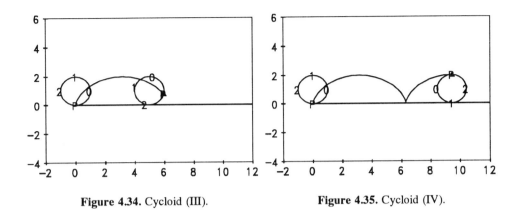

Figure 4.34. Cycloid (III).   **Figure 4.35.** Cycloid (IV).

The angle step size increment, θ_0, through which C advances with each recalculation (here, 15°), is set in Cell D3. The radius a and the distance b of P from the center of the circle are entered in Cells D4..D5. For a cycloid, $a = b$. Cell G4 counts iterations, N, using the standard @IF construction. As the circle rolls, the total amount of rotation, θ, is updated in degrees as $\theta_d = N\theta_0$ (+G4*D3) in Cell I3, and in radians as θ_r in Cell I4 (+I3*D9).

Cells H6..I6 and H7..I7 calculate the circle's center (X,Y) and the location of P, respectively, by

$$(X,Y) = (a\theta,a), \; P = (X - b\sin(\theta), Y - b\cos(\theta))$$

The center's coordinates are obtained by observing that as the circle of radius a rolls through an angle θ, it proceeds $a\theta$ units in the x-direction, while its center remains a units above the x-axis. To compute P, observe that P starts at the bottom of the circle, with an initial position angle of $-\pi/2$. After the circle has rolled through an angle θ, P has rotated clockwise about the circle's center through an angle θ, so that its coordinates become

$$(X + b\cos(-\pi/2 - \theta), Y + b\sin(-\pi/2 - \theta))$$

From basic trigonometric identities, this expression is equivalent to that above.

Columns E..F of Rows 12..36 create the original circle using 24 points, as $x = a\cos(\theta)$, $y = a + a\sin(\theta)$, $0 \leq \theta \leq 2\pi$. Column B provides a point counter, n, $0 \leq n \leq 24$, and the block B12..F36 is named T. With each recalculation the coordinates of the current, updated, locations of points on the moving circle are created in Cells B38..G62. The progress of the points on the circle is calculated by iteratively updating the counter in Column B modulo 24, as

$$m = (120 + n - N) \text{ modulo } 24$$

Each recalculation changes the iteration counter N, and thus m, by 1 to generate the rotation of P clockwise around the circle. The use of 120 in the formula insures that a positive value will result for up to five complete rotations of the circle.

The x-coordinates of the current circle are determined in Column E by adding the current center's X (i.e., H$6) to the x-value for position m found in table T. The y-coordinates are found similarly in Column G. The given formulas are entered in Row 38 and copied through Row 62.

134   *Cycloids as Rolling Circles*

Next, the current location of P is reproduced from Row 7 into Row 64. Rows 66..120 create the continually updated cycloid in Columns E and H. The given formulas are entered as indicated in Rows 66 and 67 and copied. Each of these formulas returns @NA until the value of current θ_d in Cell I3 equals the value of θ in Column C. At that time, the (x,y)-coordinates from Row 64 are reproduced in Columns E and H of that row. In subsequent recalculations these values remain unchanged.

Rows 122..123 create the graph for the *x*-axis as the base line along which the circle rolls. Column K is used for interior labels to show the progress of selected points on the circle, in order to further increase the illusion of movement. The symbol P indicates the location of the chosen point, while 0, 1, 2 are shown at 90° increments.

```
        B       C        D      E      F      G      H      I      J      K
  1  CYCLOID via Rolling Circle
  2  Parameters:            Variables:
  3    Step:θ0=    15       Start>    1   θ:d=    30
  4    Rad:a=       1         N =     2   θ:r=  0.52
  5    Dist:b=      1                         x      y
  6                        Current ctr:0.52   1.00
  7                        Current pt:  0.02  0.13
  8  Constants:
  9    pi/180=0.017         Base   Cur         Base   Cur
 10  Base Circle:           Circ   Circ  Cycl  Line   Pt
 11     n   θ:deg  θ:rad  x     y1    y2    y3    y4    y5   Lab
 12     0     0   0.00   1.00  1.00                             0
 13     1    15   0.26   0.97  1.26
  :
 30    18   270   4.71   0.00  0.00                             P
  :
 36    24   360   6.28   1.00  1.00
 37  Current Circle:
 38    22         1.39         0.50                             0
 39    23         1.49         0.74
 40     0         1.52         1.00
  :
 56    16         0.02         0.13                             P
  :
 62    22         1.39         0.50
 63
 64  Current point: 0.02                       0.13
 65  Cycloid to date:
 66      0        0.00                0.00
 67     15        0.00                0.03
 68     30        0.02                0.13
 69     45         NA                  NA
  :
120    810         NA                  NA
121  Base Line:
122                   0                  0
123                  14                  0

I3:  +G4*D3    I4:  +I3*D9           G4:  @IF(G3=0,0,1+G4)
H6:  +D4*I4                          H7:  +H6-D5*@SIN(I4)
I6:  +D4                             I7:  +I6-D5*@COS(I4)
D9:  @PI/180                         B12: 0    B13: 1+B12
C12: +B12*D$3                        D12: +C12*D$9
E12: +D$4*@COS(D12)                  F12: +D$4*@SIN(D12)+D$4
B38: @MOD(120+B12-G$4,24)
E38: @VLOOKUP(B38,$T,3)+H$6    G38: @VLOOKUP(B38,$T,4)
E64: +H7   J64: +I7                  C66: 0
E66: @IF(C66=I$3,E64,E66)      H66: @IF(C66=I$3,J64,H66)
C67: 15+C66    E67: @IF(G$3=0,@NA,@IF(C67=I$3,E$64,E67))
H67: @IF(G$3=0,@NA,@IF(C67=I$3,J$64,H67))
```

Figure 4.36. Cycloid via Rolling Circle.

Copy:		Graph:			
From	To	Series	Cells	Labels	Purpose
B13	B14..B36	X	E12..E123		x-axis
C12..F12	C13..F36	1	F12..F36	K12..K36	base circle
B38..G38	B39..G62	2	G12..G62	K12..K56	rolling circle
C67..H67	C68..H120	3	H12..H120		cycloid
		4	I12..I123		base line
		5	J12..J64		rolling point

4.5.B. Prolate and Curtate Cycloids

The curve that results from using a fixed point located in the interior of circle, is produced by setting $b < a$ to generate a *curtate cycloid* (Figure 4.37). Setting $b > a$ forms a *prolate cycloid* (Figure 4.38), the path of a point exterior to the rolling circle (as on the edge of a train wheel).

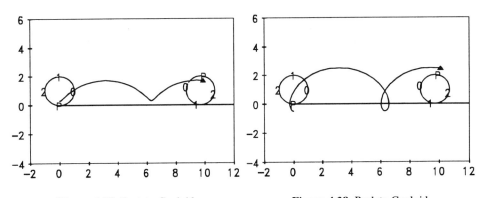

Figure 4.37. Curtate Cycloid.   **Figure 4.38.** Prolate Cycloid.

Activities

1. Expand the model with insertion and copy commands. Then use a smaller step size θ_0 for θ together with more points to generate smoother curves.
2. Explore the effects that result from using various ratios $a:b$ for the parameters a and b.
3. Replace the lookup functions in Rows 38..62 with direct computations using trigonometric functions.
4. Create animated models of the paths traced by points on other closed curves that roll over a variety of surfaces (Hall and Wagon, 1992). The topic is also mentioned in Bolt (1991), pp. 107 to 119.

4.6. HYPOCYCLOIDS AND EPICYCLOIDS

This section presents animated models that iteratively generate the curve that is formed as the path of a point P on the circumference of a circle that rolls either on the inside (*hypocycloid*) or the outside (*epicycloid*) of a given fixed circle. These curves are discussed in Bruce et al. (1990), pp. 114 to 118; Lockwood (1961), pp. 139 to 151; Yates (1974), pp. 81 to 85; Hall (1992); and Teixeira (1971), pp. 155 to 211 of Tome II. The prefix "hypo" means "beneath", while "epi" means "upon".

4.6.A. Hypocycloid

The model of Figure 4.44 creates a hypocycloid. The inner circle rolls counterclockwise around the inside of the fixed outer circle. The inner circle itself rotates clockwise about its center. The radii a of the fixed outer circle and b of the inner rolling circle, with $b < a$, are entered into Cells D6..D7. The fixed circle is created in Rows 12..36 using the standard technique in a condensed format with a step size of 15°. Initially (Figure 4.39), the inner circle is tangent to the outer circle at the point P = $(a,0)$, with its center at B = $(a-b,0)$.

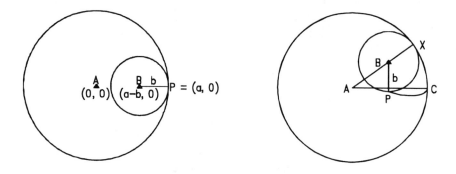

Figure 4.39. Hypocycloid: Initial.   **Figure 4.40.** Hypocycloid: Subsequent.

The model is initialized by setting Cell D3 to 0. After Cell D3 is set to a nonzero value, and with each subsequent recalculation, each position angle θ on the inner circle is incremented in steps of 15°, causing the circle to roll through an additional 15°. Illustrative output is shown in Figures 4.41 to 4.43. Cell E4 maintains the total angle rolled, $θ_d$ in degrees, as @IF(D3=0,0,15+E4). Thus, $θ_d$ is incremented by 15° with each recalculation. The radian equivalent $θ_r$ is calculated in Cell F4 as +E4*D9, where D9 computes π/180.

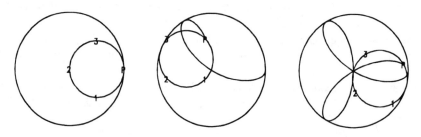

Figure 4.41. Hypocycloid.   **Figure 4.42.** Hypocycloid.   **Figure 4.43.** Hypocycloid.

Figure 4.40 helps to locate the circle's center. The point X represents the current point of tangency. After the inner circle has rolled so that ∡PBX = $θ_r$, the length of the arc PX is $θ_r b$. If ∡XAC = α is the corresponding angle in the fixed circle, then $αa = θ_r b$, so that

$\alpha = (b/a)\theta_r$. Since the center of the inner circle is $a-b$ units from the origin, its center B, (x_0, y_0), given in Cells F6,G6, is $x_0 = (a-b)\cos((b/a)\theta_r)$, $y_0 = (a-b)\sin((b/a)\theta_r)$.

```
           B         C         D         E         F         G
 1 HYPOCYCLOID via Internal Circle
 2 Start:
 3    START >            1 θd:deg  θr:rad
 4                         30      0.524
 5 Curve Radii:                     x        y
 6   Outer, a=           1 Center: 0.483   0.1294
 7   Inner, b =   0.5
 8 Constant:
 9   pi/180 =    0.017
10                 Fixed    Inner            Curve
11 θ(deg)    x      y1       y2     Label    y3
12    0     1.00   0.00
13   15     0.97   0.26
14   30     0.87   0.50
 :
36   360    1.00  -0.00
37 Moving Circle:
38   30     0.92           -0.12      P
39   45     0.84           -0.22
40   60     0.73           -0.30
 :
44   120    0.23           -0.30      1
 :
50   210    0.05            0.38      2
 :
56   300    0.73            0.56      3
 :
62    30    0.92           -0.12
63 Hypocycloid:
64     0    1.00                              0.00
65    15    0.98                             -0.06
66    30    0.92                             -0.12
67    45     NA                                NA
 :
183 1785    NA                                 NA

E4: @IF(D3=0,0,15+E4)        F4: +E4*D9
F6: (D6-D7)*@COS(F4*D7/D6)
G6: (D6-D7)*@SIN(F4*D7/D6)
D9: @PI/180                  B12: 0   B13: +B12+15
C12: +D$6*@COS(B12*D$9)      D12: +D$6*@SIN(B12*D$9)
B38: @MOD(B12+E$4,360)
C38: +F$6+D$7*@COS(-B38*D$9)
E38: +G$6+D$7*@SIN(-B38*D$9)
B64: 0     B65: 15+B64
C64: +D6   C65: @IF(D$3=0,@NA,@IF(B65=E$4,C$38,C65))
G64: 0     G65: @IF(D$3=0,@NA,@IF(B65=E$4,E$38,G65))
```

Figure 4.44. Hypocycloid.

Copy:		Graph:			
From	To	Series	Cells	Labels	Purpose
B13	B14..B36	X	C12..C183		x-axis
C12..D12	C13..D36	1	D12..D36		fixed circle
B38..E38	B39..E62	2	E12..E62	F12..F62	rolling circle
B65..G65	B66..G183	3	G12..G183		hypocycloid

138 *Hypocycloids and Epicycloids*

Rows 38..62 generate the current coordinates of the rolling circle. When $\theta = \theta_d$, a point on its circumference has proceeded through an angle of $-\theta_d$ relative to the center of the inner circle. Thus, to update angles on the moving circle in steps of 15°, @MOD(B12+E$4,360) is first entered in Cell B38 and then copied into Cells B39..B62. This generates the angle for each point on the inner circle. Their (x,y)-coordinates are created in Columns C and E by

$$x = x_0 + b\cos(-\theta), \quad y = y_0 + b\sin(-\theta)$$

with θ converted to radians. The formulas are entered in Row 38 and then copied through Row 62. With this scheme, the top row of the block (Row 38) always contains the current location of the moving point P.

Finally, the hypocycloid is created in Rows 64..183 through a series of recalculations. First, the initial point $(a,0)$ is reproduced in Row 64. The remaining formulas initially return @NA for the coordinates. These remain fixed until θ_d (Cell E4) reaches the θ value in Column B, when the current point from Row 38 is reproduced in that row. They then remain unchanged in ensuing recalculations.

Once the model has been constructed, it is instructive to use it to observe the effects of changing the ratio between the radii a and b. For certain ratios the model produces classical "roses" with varying numbers of leaves, while other ratios require several passes around the circle before a closed curve is formed. For some choices of a and b it is beneficial to extend the number of rows in the hypocycloid section of the graph by the insert and copy technique.

4.6.B. Epicycloid

It is easy to adjust the model for a circle that rolls counterclockwise on the outside of a fixed circle to form an epicycloid. One way to do this is to change three pairs of formulas in Figure 4.44 and reverse the roles of a and b. In Cells F6..G6 the center of the rolling circle (x_0, y_0) becomes

$$x_0 = (a+b)\cos((b/a)\theta_r), \quad y_0 = (a+b)\sin((b/a)\theta_r)$$

Also, the coordinates of the point P are

$$x = x_0 + b\cos(\theta), \quad y = y_0 + b\sin(\theta)$$

because the outer curve rotates in a clockwise direction relative to its center. However, now point 2 is traced. The formulas to be changed are

```
F6:  (D6+D7)*@COS(F4*D7/D6)        G7:  (D6+D7)*@SIN(F4*D7/D6)
C38: +F$6+D$7*@COS(B38*D$9)        E38: +G$6+D$7*@SIN(B38*D$9)
C65: @IF(..,...,@IF(..,C$50,..))   G65: @IF(..,...,@IF(..,E$50,..))
```

with the second pair copied into Rows 39..62 and the last pair copied into Rows 66..183. Some illustrative graphs that are produced are shown in Figures 4.45 and 4.46. Epicyclic gear trains are discussed in Bolt (1991), pp. 25 to 39.

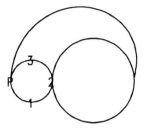
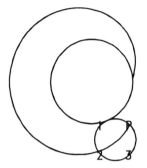

Figure 4.45. Epicycloid (I). **Figure 4.46.** Epicycloid (II).

Activities

1. Investigate the hypocycloid model using various choices of b for a fixed value a.
2. Create an animated epicycloid model. Experiment with various choices of a and b. Use the model to investigate how one coin rolls around another.
3. Use the insert and copy commands to modify the models to include more points in order to produce smoother curves. Modify the models so that the angle increments are parameters, or so that the circles roll in the opposite direction.
4. Modify the hypocycloid and epicycloid models to trace the progress of a point on a radius c units from the center of the rolling circle, thus obtaining animated models for the hypotrochoid and epitrochoid curves of Section 2.13.
5. What change is produced in the hypocycloid when b is replaced by $a-b$?

4.7. INVOLUTE OF A CIRCLE

Suppose that a string, originally wrapped around a circle C, is unwound so that it remains taut and tangent to C. The curve traced by its endpoint P is the *involute* of the circle. This construction is examined in Yates (1974), pp. 135 to 137, and is mentioned in most calculus books. As shown in the graphs of Figures 4.47 and 4.48, the model of Figure 4.49 produces the curve sequentially.

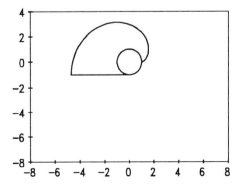
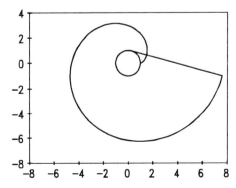

Figure 4.47. Involute of Circle (I). **Figure 4.48.** Involute of Circle (II).

140   *Involute of a Circle*

```
        B     C      D E     F       G    H    I
 1  INVOLUTE OF CIRCLE
 2  Parameters:
 3    Step:θ0=     15   Start>    1
 4
 5  Curve Parameters:
 6         a =    1       α(d)=   30
 7                        α(r)= 0.52
 8  Constants:
 9    pi/180=0.0175
10                                 Circ Line Invo
11     n  θ:rad  θ:deg     x      y1   y2   y3
12     0    0    0.00    1.00   0.00
13     1   15    0.26    0.97   0.26
14     2   30    0.52    0.87   0.50
       :
36    24  360    6.28    1.00  -0.00
37
38  Line: tan.point       0.87         0.50
39        end point       1.13         0.05
40
41  Invo:  0    0.00     1.00              0.00
42         15   0.26     1.03              0.01
43         30   0.52     1.13              0.05
44         45   0.79      NA                NA
       :
159      1770  30.89      NA                NA

G6:   @IF(G3=0,0,D3+G6)   G7:  +G6*D9
D9:   @PI/180              B12: 0      B13: 1+B12
C12:  +B12*D$3             D12: +C12*D$9
F12:  +D$6*@COS(D12)       G12: +D$6*@SIN(D12)
F38:  +D6*@COS(G7)         H38: +D6*@SIN(G7)
F39:  +D6*(@COS(G7)+G7*@SIN(G7))
H39:  +D6*(@SIN(G7)-G7*@COS(G7))
C41:  0                    C42: 15+C41
D41:  +D9*C41              F41: +D6    I41: 0
F42:  @IF(G$3=0,@NA,@IF(C42=G$6,F$39,F42))
I42:  @IF(G$3=0,@NA,@IF(C42=G$6,H$39,I42))
```

Figure 4.49. Involute of Circle.

Copy:		Graph:			
From	To	Series	Cells	Labels	Purpose
B13	B14..B36	X	F12..F159		x-axis
C12..G12	C13..G36	1	G12..G36		circle
C42..I42	C43..I159	2	H12..H39		string
D41	D42..D159	3	I12..I159		involute

The circle's radius a is entered in Cell D6. Cell G3 is used to initialize the model. When Cell G3 is set to 0, none of the string is unwound. With each recalculation after Cell G3 is set to a nonzero value, the angle α (in degrees) in Cell G6 that gives the string's point of tangency P_1 is increased in increments of $\alpha = 15°$ (set in Cell D3) by the formula @IF(G3=0,0,D3+G6). The corresponding radian measure is determined in Cell G7. The circle is created in Cells B12..G36 by the standard scheme with $x = a\cos(\theta)$, $y = a\sin(\theta)$.

From Figure 4.50 (where t has been used for α) the point of tangency of the string is $(a\cos(\alpha), a\sin(\alpha))$. Because $[\sin(\alpha),-\cos(\alpha)]$ is a unit tangent vector, and the length of the unwound string is $a\alpha$, then the string vector is $a\alpha[\sin(\alpha),-\cos(\alpha)]$. Thus, in Rows 38 and

39 the coordinates of the string's tangent point and endpoint are computed as

[38]: $x = a\cos(\alpha)$, $y = a\sin(\alpha)$, [39]: $x = a(\cos(\alpha)+\alpha\sin(\alpha))$, $y = a(\sin(\alpha)-\alpha\cos(\alpha))$

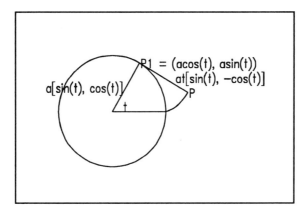

Figure 4.50. Formula: Involute of Circle.

Finally, in Rows 41..159 the endpoint of Row 39 is used to generate the involute. Row 41 reproduces the base point $(a,0)$. To build the remaining rows, the indicated formulas are entered in Row 42 and copied through Row 159. Each of these initially returns @NA. During recalculation, when the current value of α (Cell G6) matches the θ in Column C, the (x,y)-values are read into that row from Row 39; otherwise, each cell remains unchanged. Thus, the graph of the involute grows by a new point with each recalculation.

Activities

1. Modify the model by decreasing the size of the step increment θ_0 and generating more points to produce smoother curves.
2. Suppose that a length of string is wound around a system of two circles, each of radius a, whose centers are b units apart, with $b \geq 2a$. The string is unwound so that it is held taut and remains tangent to the current circle of contact (see Figures 4.51 and 4.52). Create an animated model that generates the path of the end of the string that results as it is unwound. As a further activity, generalize the model to the case in which the circles have different radii.

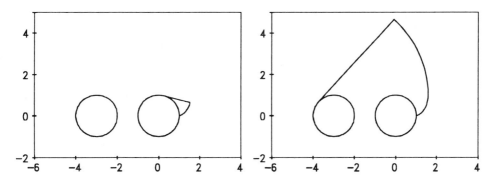

Figure 4.51. String, Stage I.   **Figure 4.52.** String, Stage II.

142 *Glissettes*

4.8. GLISSETTES

If one curve slides against one or more fixed curves, the path traced by a point on the sliding curve is called a *glissette*. This topic is pursued in Lockwood (1961), pp. 160 to 165, and Yates (1974), pp. 108 to 112. The graphs of Figures 4.53 and 4.55 were created from Figure 4.57 to show a rectangle sliding against the positive *x*- and *y*-axes to generate curves traced by two points P_1 and P_2. In this example, P_1 is on the left edge of the rectangle, while P_2 lies in its interior.

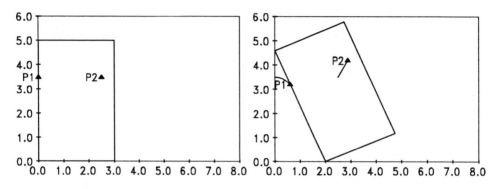

Figure 4.53. Glissette (I). **Figure 4.54.** Glissette (II).

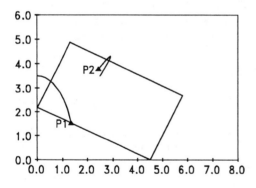

Figure 4.55. Glissette (III).

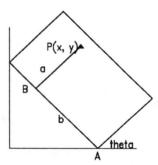

Figure 4.56. Glissette Formulas.

Cell D1 is used to initialize the process. When Cell D1 is set to 0, the rectangle is in its original upright position, with its lower left corner at the origin (Figure 4.53). After Cell D1 is set to a nonzero value, with each recalculation the lower left corner of the rectangle slides an additional *d* units along the positive *x*-axis.

As the rectangle slides, P_1 traces out part of an ellipse, always moving to a lower position, while P_2 initially moves upward before descending. It is easy to produce the paths of other points by changing P_1 and P_2. Supplying the additional rows that draw the rectangle and other points are left as activities.

The dimensions H and W of the rectangle are set in Cells D4..D5, with the initial coordinates (a_i, b_i) of the points P_1 and P_2 entered in Cells F4..G4 and F5..G5, and the *x*-step size *d* placed in Cell D3. Here, $d = 0.5$. Cell F1 counts iterations, *N*. The current *x*-coordinate, x_0, of the lower left corner of the rectangle is calculated in Cell D7 as *Nd* (or

+F1*D3), while Cell D8 provides the angle of the base of the rectangle as $\theta = \arcsin(x/H)$.

Cells F7,G7 and F8,G8 keep track of the current coordinates (x,y) of the points P_1,P_2. The formulas are derived using Figure 4.56 with vector addition to find the coordinates of the points A, B, and P successively:

A: $(x_0,0)$
B: $(x_0,0) + b[\cos(\theta+\pi/2),\sin(\theta+\pi/2)]$
P: $(x_0+b\cos(\theta+\pi/2),b\sin(\theta+\pi/2)) + a[\cos(\theta),\sin(\theta)]$
P: $(x_0+b\cos(\theta+\pi/2)+a\cos(\theta),b\sin(\theta+\pi/2)+a\sin(\theta))$

Trigonometric identities can be used to simplify the expression for the coordinates of P.

The path of the first curve is traced in Columns C..D of Rows 12..32. Column B serves as an iteration counter, n. Initially the (x,y)-coordinates of these rows are @NA. When the value of N in Cell F2 reaches that of the n in Column B, then the values from Row 7 are reproduced into the row given by n. At all other times these coordinates remain unchanged. A similar process is used to form the second path in Rows 34..54 using the values in Row 8. With each recalculation the curves grow by one point each. Rows 55..56 generate a line that shows the progress of the left edge of the rectangle.

```
         B        C       D      E       F       G
 1              START>    1     N =      2
 2
 3     Step, d=          0.5  Points     a       b
 4     RecHt,H=           5    1st>      0      3.5
 5     RecWt,W=           3    2nd>     2.5     3.5
 6                             Points    x       y
 7     LCnr,x0=           1    1st>    0.30    3.43
 8     Angle:θ=         0.201  2nd>    2.75    3.93
 9
10  Path 1:         Path1  Path 2   Rect  Point
11     n       x     y1     y2      y3     y4
12     0     0.00   3.50
13     1     0.15   3.48
14     2     0.30   3.43
15     3      NA     NA
 :
32    20      NA     NA
33  Path 2:
34     0     2.50          3.50
35     1     2.64          3.73
36     2     2.75          3.93
37     3      NA            NA
 :
54    20      NA            NA
55    Edge:   1                     0
56            0                    4.9

F1: @IF(D1=0,0,1+F1)
D7: +F1*D3      D8: @ASIN(D7/D4)
F7: +D$7+F4*@COS(D$8)+G4*@COS(D$8+@PI/2)
G7: +F4*@SIN(D$8)+G4*@SIN(D$8+@PI/2)
F8: +D$7+F5*@COS(D$8)+G5*@COS(D$8+@PI/2)
G8: +F5*@SIN(D$8)+G5*@SIN(D$8+@PI/2)
B12: 0     B13: 1+B12    C12: +F4    D12: +G4
C13: @IF(D$1=0,@NA,@IF(B13=F$1,F$7,C13))
D13: @IF(D$1=0,@NA,@IF(B13=F$1,G$7,D13))
B34: +B12    C34: +F5    E34: +G5
C35: @IF(D$1=0,@NA,@IF(B35=F$1,F$8,C35))
E35: @IF(D$1=0,@NA,@IF(B35=F$1,G$8,E35))
C55: +D3*F1      F55: 0
C56: 0           F56: @SQRT(D4^2-C55^2)
```

Figure 4.57. Glissette (Sliding Rectangle).

144   *Glissettes*

Copy:		Graph:			
From	To	Series	Cells	Labels	Purpose
B13..D13	B14..D32	X	C12..C54		x-axis
B34	B35..B54	1	D12..D32		path 1
C35..E35	C36..E54	2	E12..E54		path 2
		3	F12..F56		line

Activities

1. Vary P_1 and P_2 to find the paths produced by other points within the rectangle.
2. Supplement the model to draw the entire sliding rectangle, and the current locations of points P_1 and P_2. Simplify the formulas in Rows 7..8.
3. Create glissettes for other curves, such as those generated by a point on a line segment whose endpoints move on (a) two lines intersecting at a nonright angle; (b) two circles of equal radii; (c) a parabola and its directrix.
4. Four bugs start at the corners of a square table. Each moves at a constant rate in the direction of the bug on its right. Create an animated model to trace the bugs' paths. (See Steinhaus, 1969, p. 136; Lockwood, 1961, p. 99.)
5. Modify the model of Figure 4.57 to use only the line segment. Then cause the line segment to continue to move so that one end remains on the *x*-axis and the other on the *y*-axis until the entire ellipse is traced out.

4.9. STRAIGHT-EDGE AND TRIANGLE CONSTRUCTIONS

Many curves can be sketched by using a straight-edge and a right triangle. The model of this section creates a parabola as an envelope of lines. (See Lockwood, 1961, pp. 2 to 3; Pedoe, 1976, pp. 202 to 203.) It first draws a fixed line L and a point A not on L. Here, A is (2,-2) and L is the *y*-axis. The triangle's short leg is placed on A and its right angle on L. A line then is drawn along the long leg. With each recalculation the model moves the triangle's right angle down L and draws another line, creating Figures 4.58 and 4.59.

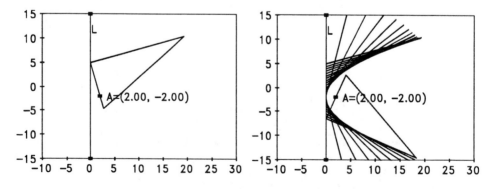

Figure 4.58. Parabola via Triangle.   **Figure 4.59.** Parabola via Triangle.

In Figure 4.60 the lengths, d_l and d_s, of the triangle's legs are entered in Cells D7 and D8, with A = (x_a, y_a) set in Cells F7..F8, and the *y*-values of high and low points for L put in Cells I7..I8. The step size, h, for incremental changes in y is set in Cell D3, with the initial of y, y_0, entered in Cell D4. Cell H3 is used to initialize the process, while Cell H4 serves as an iteration counter N. When Cell H3 is set to 0, the triangle is placed in an initial position above the fixed point and a line drawn. After Cell H3 is set to a nonzero value, and with each subsequent calculation, the triangle moves slightly, and another line is drawn.

```
        B       C      D       E       F       G       H       I
 2  Parameters:        PARABOLA: Envelope via Triangles
 3      Step:h=    0.5         START>          1
 4      Init:y0=   5           N =             2
 5
 6  Triangle:          Point A:               Line:
 7      Long Leg: 20   xa =     2             Lo,y=   -15
 8      Short Leg:10   ya =    -2             Hi,y=    15
 9
10  Graphs:            line/pt  tri  lines    slope  angle
11              x      y1       y2   y3       m        α
12      Line:   0.00  -15.00
13              0.00   15.00
14                     NA        NA
15      Point:  2.00   -2.00
16                                             slope  angle
17      Triang: 0.00            4.00   Shrt   -3.00  -1.25
18              3.16           -5.49   Long    0.33   0.32
19             18.97           10.32
20              0.00            4.00
21          n   mod
22          0    0    0.00             5.00
23          0    1   19.23            10.49
24          0    2    NA               NA
25          1    0    0.00             4.50
26          1    1   19.12            10.38
27          1    2    NA               NA
28          2    0    NA               NA
     :
105        27    2    NA               NA

H4:  @IF(H3=0,0,1+H4)          D12: 0     E12: +I7
D13: 0    E13: +I8             D15: +F7   E15: +F8
D17: 0                         F17: @IF(H3=0,D4,F17-D3)
D18: +D17+D8*@COS(I17)         F18: +F17+D8*@SIN(I17)
D19: +D17+D7*@COS(I18)         F19: +F17+D7*@SIN(I18)
D20: +D17                      F20: +F17
H17: (F17-F8)/(D17-F7)         I17: @ATAN(H17)
H18: -1/H17                    I18: @ATAN(H18)
B22: 0                         B23: +B22+(C23=0)
C22: 0                         C23: @MOD(1+C22,3)
D22: @IF(B22>H$4,@NA,@IF(B22=H$4,
                  @CHOOSE(C22,D$20,D$19,@NA),D22))
G22: @IF(B22>H$4,@NA,@IF(B22=H$4,
                  @CHOOSE(C22,F$20,F$19,@NA),G22))
```

Figure 4.60. Straight-Edge and Triangle.

Copy:		Graph:			
From	To	Series	Cells	Labels	Purpose
B23..C23	B24..C105	X	D12..D105		x-axis
D22..G22	D23..G105	1	E12..E15		line/point
		2	F12..F20		triangle
		3	G12..G105		line family

The depictions of L and the point A are created in Rows 12..15. The moving triangle is generated by plotting its three vertices in Rows 17..19, with the initial vertex repeated in Row 20 to complete the triangle. Cell F17 determines the current y-coordinate of the right angle by @IF(H3=0,D4,F17-D3). Thus, initially the right angle is at y_0 (i.e., D4), and with each recalculation it is decreased by h (i.e., D3) units. If (u_0,v_0) denotes the right angle, then the slope and angle of the short leg are found in Cells H17 and I17 as

$$m_1 = (v_0-y_0)/(u_0-x_0) \text{ and } \alpha_1 = \arctan(m_1)$$

The slope and angle of the long leg are given by $m_2 = -1/m_1$ and $\alpha_2 = \arctan(m_2)$ in Cells H18..I18. The other vertices, are formed in Rows 18 and 19 by

$$[18]: (u_1,v_1) = (u_0,v_0) + d_1[\cos(\alpha_1),\sin(\alpha_1)]$$

$$[19]: (u_2,v_2) = (u_0,v_0) + d_2[\cos(\alpha_2),\sin(\alpha_2)]$$

Rows 22..105 use the triangle to draw the lines. The scheme is an elaboration of the multiple line technique of Section 1.6. When Cell H3 is 0, the (x,y)-coordinates of all but those of the first line (Rows 22..23) are @NA. After Cell H3 is set to a nonzero value, these remain unchanged until the row counter n in Column B equals the iteration counter N, at which time the @CHOOSE function shown reproduces the coordinates of the end of the current long leg from Rows 19 and 20 into the two rows for that value of n. These then remain unchanged in subsequent recalculations. One new line is produced with each recalculation.

Activities

1. Create an animated model for the following construction. Draw a circle C with center O and diameter UOV. Mark a point A on UV. Using a right triangle, from any point P on C draw a chord PQ at right angles to PA. Repeat this operation as P moves around C with A fixed. The envelope of the lines is an ellipse.
2. Illustrate some of the envelope models of Chapter 2 via a moving triangle.
3. See Yates (1974) and Lockwood (1961) for examples constructed with a right triangle. Create a model for Newton's method of constructing cissoids and strophoids using a carpenter's square (Yates, 1974, pp. 28 to 29).

4.10. LINKAGES

Many curves arise as paths that are traced by mechanisms called *linkages*. Often it is possible to illustrate the operation of a linkage through an animated spreadsheet model. Figure 4.61 illustrates one example (Lockwood, 1961, pp. 114 to 116). Three rods of lengths 1, $\sqrt{2}$, and 1 are connected by hinges (open squares). Two ends (solid squares) are affixed at the points (0,0) and ($\sqrt{2}$,0) so that the rods rotate about them. If a pen is placed at the midpoint of the center rod (triangle), a lemniscate is traced out as the system moves.

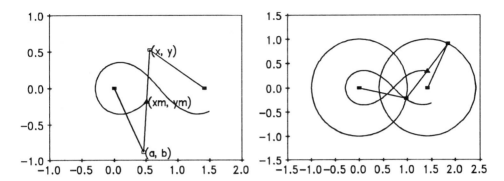

Figure 4.61. Linkage (Lemniscate). **Figure 4.62.** Linkage II.

The animated model, presented in Figure 4.63, is initialized in Cell I1. When Cell I1 is set to 0, the rods are located in an initial position. After Cell I1 is set to a nonzero value, and with each subsequent recalculation, the left hinge rotates about the origin, generating movement of the linkage system and tracing a section of the curve.

The left rod is rotated about a pivot point iteratively in steps of $\theta_0 = 2.5°$ (set in Cell D3). That rod is rotated through k (set in Cell I4) steps of θ_0 in each recalculation. Cell I5 counts N, the total number of steps to date, by @IF(I1=0,0,I5+I4). In Cell D5 the relative location p, $0 \le p \le 1$, of the pen on the center rod is set. Here, $p = 0.5$. To avoid computational difficulties that can arise in finding the square root of a number near 0, in Column C values of θ are generated by including an offset of θ_1 (Cell D4) added to the usual multiples of θ_0.

Because of its large size, this model is displayed in two sections in Figure 4.63. Column B counts points. Column C computes the angles θ in degrees, starting with $\theta = \theta_1$ in increments of θ_0, each offset by the amount D$4. Thus, the formula for θ in Cell C12 is +D$3*B12+D$4. Column D converts the angles to radians. The moving hinge (a,b) of the left rod traces a circle of radius 1 about the fixed point $(0,0)$. Points on this circle are created in Columns F and G as $(\cos(\theta),\sin(\theta))$. Once the left hinge (a,b) is located, the coordinates of the other hinge (x,y) are determined. The squares of the lengths of the two arms ending at (x,y) satisfy $(x-a)^2 + (y-b)^2 = 2$, $(x-\sqrt{2})^2 + y^2 = 1$. After expansion, these become

$$x^2 - 2ax + a^2 + y^2 - 2by + b^2 = 2$$
$$x^2 - 2\sqrt{2}x + 2 + y^2 \qquad\qquad = 1$$

Using the fact that $a^2 + b^2 = 1$, subtracting, and solving for y yields

$$(1) \qquad y = ((-a+\sqrt{2})x - 1)/b$$

Substituting Equation (1) into $x^2 - 2\sqrt{2}x + 2 + y^2 = 1$ produces $a_2x^2 + a_1x + a_0 = 0$, where

$$a_2 = 1+(\sqrt{2}-a)^2/b^2, \quad a_1 = -2\sqrt{2}-2(\sqrt{2}-a)/b^2, \quad a_0 = 1+1/b^2$$

This can be solved for x by using the quadratic formula, with the negative root selected to find the appropriate location using Equation (2). Columns I..K determine a_2, a_1, a_0. Columns M and N find the coordinates of the hinge (x,y) from Equations (1) and (2).

$$(2) \quad x = \frac{-a_1 - \sqrt{a_1^2 - 4a_2 a_0}}{2a_2}$$

Columns O and P determine the intermediate point (X_m, Y_m) that generates the curve by using $p = \frac{1}{2}$ in the relation $X_m = (1-p)a + px$, $Y_m = (1-p)b + py$.

The path of the linkage is generated in Columns R..S using the usual point-by-point method, with one new point added with each recalculation. To create the model, the indicated formulas are entered in Row 12, and copied through Row 158. The range B12..S158 is named as the table T. Finally, the formulas in Rows 160..164 use T to generate the moveable rods, hinges, and other facets of the model.

```
         B    C      D    E  F    G    H    I     J      K     L  M    N    O    P
  1 LINKAGE              Start>   0
  2 Parameters:
  3   Step:θ0=  2.5      Plot:
  4   Offset=10.05         k =    5
  5   Pen:p=    0.5        N =    0
  :
  9      pi/180=0.02
 10 Curve:                                              Hinge      Midpt
 11   n    θ:d   θ:r    a     b    a2    a1    a0     Xh    Yh    Xm    Ym
 12   0   10.1  0.18 : 1.0   0.2 : 7.1  -31.0  33.8 : 2.0  -0.8   1.5  -0.3
 13   1   12.6  0.22 : 1.0   0.2 : 5.1  -21.4  22.2 : 1.8  -0.9   1.4  -0.3
  :
157 145  372.6  6.50 : 1.0   0.2 : 5.1  -21.4  22.2 : 1.8  -0.9   1.4  -0.3
158 146  375.1  6.55 : 1.0   0.3 : 4.0  -16.1  15.8 : 1.7  -1.0   1.3  -0.4


I5:   @IF(I1=0,0,I5+I4)    D9: @PI/180       B12: 0          B13: 1+B12
C12:  +D$3*B12+D$4         D12: +C12*D$9     F12: @COS(D12)  G12: @SIN(D12)
I12:  1+(@SQRT(2)-F12)^2/G12^2    J12: -2*@SQRT(2)-2*(@SQRT(2)-F12)/G12^2
K12:  1+1/G12^2            M12: (-J12-@SQRT(J12*J12-4*I12*K12))/(2*I12)
N12:  ((-F12+@SQRT(2))*M12-1)/G12      O12: (1-D$5)*F12+D$5*M12
P12:  (1-D$5)*G12+D$5*N12
R12:  +O12                 R13: @IF(I$1=0,@NA,@IF(I$5>=B13,O13,R13))
S12:  +P12                 S13: @IF(I$1=0,@NA,@IF(I$5>=B13,P13,S13))
R160: 0                    R161: @VLOOKUP(I5,T,4)   R162: @VLOOKUP(I5,T,11)
R163: @SQRT(2)             R164: (1-D5)*R161+D5*R162
T160: 0                    T161: @VLOOKUP(I5,T,5)   T162: @VLOOKUP(I5,T,12)
T163: 0                    U164: (1-D5)*T161+D5*T162
V161: +T161                V162: +T162                       W160: 0  W163: 0


         R      S      T      U      V      W
  9                         Moving     Fixed
 10   LinkPath Arms  PointHingePoint
 11     x     y1     y2     y3     y4     y5
 12    1.5   -0.3
 13    NA    NA
  :
158    NA    NA
159
160    0.0          0.0           0.0
161    1.0          0.2           0.2
162    2.0         -0.8          -0.8
163    1.4          0.0           0.0
164    1.5                -0.3
```

Figure 4.63. Linkage (Lemniscate).

Copy:		Graph:			
From	To	Series	Cells	Labels	Purpose
B13	B14..B158	X	R12..R164		x-axis
C12..P12	C13..P158	1	S12..S158		path
R13..S13	R14..S158	2	T12..T163		rods
		3	U12..U164		moving point
		4	V12..V162		hinges
		5	W12..W163		fixed points

Activities

1. Investigate the effect of changing p, the location of the pen, in Cell D5. Also, design a graph that traces the paths of the hinges, as in Figure 4.62.
2. Find other linkage problems, and implement them in spreadsheet models. Figure 4.64 illustrates another linkage. One rod of length 1 swivels on a fixed pivot at (0,0). The other the rod is attached to the first rod on a swivel, and is allowed to slide through a fixed pivot at (-1,0). The end of the rod traces a limaçon. See also examples in Cundy and Rollett (1961), pp. 220 to 235; Bruce et al. (1990), pp. 228 to 245; Bolt (1991), pp. 57 to 94; Courant and Robbins (1941), pp. 155 to 158; and especially Yates (1974).

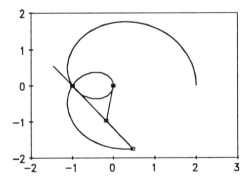

Figure 4.64. Linkage (Limaçon).

3. Modify the model of this section so that the rod lengths and the pivot locations are parameters. Some curves are shown in Brieskorn and Knörrer (1986), pp. 58 to 62.
4. Examine Bolt (1991) for numerous sources of other animation models, including cams, pulleys, chains, gear trains, and winding mechanisms.

4.11. CIRCULAR FUNCTIONS

Trigonometric, or circular, functions can be defined in terms of the progress of a point P moving counterclockwise around the unit circle, starting at (1,0). After P has advanced through an angle of θ radians, its coordinates are $(\cos(\theta), \sin(\theta))$. At this location P has traveled θ units along the circumference of the circle. Figures 4.65 to 4.66 illustrate the definition of the sine function as the y-coordinate of P. These are produced by the model

150 *Circular Functions*

of Figure 4.67. With each recalculation, P moves around the circle through angle increments of $5\pi/180$ radians (or 5°). Simultaneously, a graph showing $\sin(\theta)$ as a function of θ grows at the right. In a computer implementation these diagrams can be enhanced advantageously through the creative use of line coloring and styles.

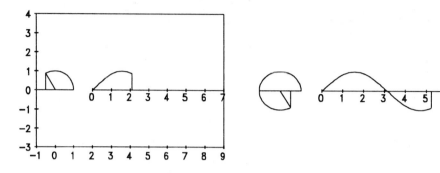

Figure 4.65. $y = \sin(\theta)$.   **Figure 4.66.** $y = \sin(\theta)$.

```
           B       C       D       E       F       G       H       I       J       K
    2           Start>     1               Current:
    3           N =        2               Deg    Rad(t) Sin(y)
    4           π/180=0.017                 10     0.17   0.17
    :
   10   Circle:                     Circ   Sine    Arc  AxisAux
   11    n     θ:d    θ:r     x      y     y1      y2    y3   y4  Labels
   12    0       0   0.00   1.00   0.00           0.00
   13    1       5   0.09   1.00   0.09           0.09
   14    2      10   0.17   0.98   0.17           0.17
   15    3      15   0.26   0.97   0.26           NA
    :
   84   72     360   6.28   1.00  -0.00           NA
   85   Moving Curve:
   86    0                         2.00   0.00
   87    1                         2.09   0.09
   88    2                         2.17   0.17
   89    3                          NA     NA
    :
  158   72                          NA     NA
  159   t-axis:                      2                    0           0
  160                                3                    0           1
    :
  166                                9                    0           7
  167          Radius:     0.00                                 0.0
  168                      0.98                                 0.2
  169                      0.985                                0.0
  170                       NA                                   NA
  171          Vert2:      2.17                                 0.2
  172                      2.17                                 0.0
  173          Base:        -1                            0
  174                        1                            0

   D3: @IF(D2=0,0,1+D3)    D4: @PI/180           F4: 5*D3
   G4: +D4*F4              H4: @SIN(G4)          B12: 0    B13: 1+B12
   C12: 5*B12              D12: +D$4*C12         E12: @COS(D12)
   F12: @SIN(D12)          H12: @IF(C12<=F$4,F12,@NA)
   B86: +B12               E86: @IF(B86>D$3,@NA,2+D12)
   G86: @IF(B86>D$3,@NA,F12)
   E159: 2      E160: 1+E159      I159: 0              K159: +E159-2
   E167: 0          J167: 0      E168: @COS(G4)        J168: @SIN(G4)
   E169: +E168      J169: 0      E170: @NA             J170: @NA
   E171: 2+G4       J171: +H4    E172: +E171           J172: 0
   E173: -1         H173: 0      E174: 1               H174: 0
```

Figure 4.67. Sketch of Growth of $y = \sin(\theta)$.

Copy:		Graph:			
From	To	Series	Cells	Labels	Purpose
B13	B14..B84	X	E12..E174		x-axis
C12..H12	C13..H84	1	G12..G158		arc of circle
B86..G86	B87..G158	2	H12..H174		sine, circle base
E160	E161..E166	3	I12..I166	K12..K166	graph axis
I159..K159	I160..K166	4	J12..J172		radius, vert. lines

Cell D2 is used to initialize the model in the usual fashion, with Cell D3 providing a count of iterations N. Cell F4 uses N to determine the current value of θ in degrees as $5N$, while Cells G4 and H4 compute θ in radians and $\sin(\theta)$, respectively.

The entire unit circle is generated in Columns E..F of Rows 12..84. It has not been included in the graph here, but this can be done especially effectively using colors. Rows 159..166 generate a line to represent the positive x-axis. To relocate this formation of the x-axis away from the circle, 2 is added to each x-coordinate. Internal labels in Column K then provide its correct coordinates.

Column H employs the animation techniques of Section 4.1 to generate the unit circle iteratively. Initially each y-value in the range H12..H84 is NA. After Cell D2 is set to a nonzero value, with each recalculation one more y-coordinate is reproduced from Column F into Column H.

The sine function is generated in Rows 86..158 in a similar point-by-point fashion by using the method of Section 4.3. However, 2 is added to each x-value in Column E in order to displace the graph away from the circle. Finally, Rows 167..174 create auxiliary lines, such as the radius and the base of the circle.

The model can be supplemented by including additional auxiliary lines and text. In addition, the border labels that are provided by the spreadsheet can be hidden, as shown in Figure 4.66.

Activities

1. Complete the model to draw the auxiliary lines. Use colors and line styles effectively.
2. Modify the model to illustrate the cosine function.
3. Create a similar model using a hyperbola, rather than a circle, to illustrate some of the hyperbolic functions (Finney and Thomas, 1990, p. 515).

4.12. CURVES OF CONSTANT WIDTH

As a circle C of radius a rolls along a line, the distance of the point on C that is furthest from the line (i.e., the diameter) remains constant. A curve with this property is called a *curve of constant width*. Perhaps surprisingly, there are many such curves in addition to circles. Figure 4.70 describes a model for one of them, exhibiting the curve's features by rotating it inside a square, as in Figure 4.68. The curve is created by starting from a equilateral triangle. From each vertex an arc of a circle is drawn passing through the other two vertices. The resulting figure is called a *Reuleaux triangle*. Curves of constant width are discussed in Yaglom and Boltyanskii (1961), pp. 70 to 82, and Bolt (1991), p. 108.

152 Curves of Constant Width

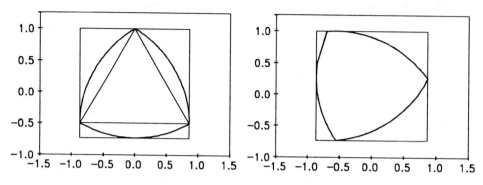

Figure 4.68. Curve of Constant Width (I).

Figure 4.69. Curve of Constant Width (II).

```
        B       C       D        E       F        G       H       I
  1  CONSTANT WIDTH
  2  Parameters:     Adjust:         Constants:
  3  Start>      0   Ax =   6E-17    √3  =  1.732
  4  α:deg=      0   Ay =  -2E-16    π/180= 0.017
  5  α:rad=      0
  6                           tri   square  base    move
  7                  x        y1    y2      y3      y4      Xr      Yr
  8   Tri:    0.00   1.00
  9          -0.87  -0.50
 10           0.87  -0.50
 11           0.00   1.00
 12  Square    NA                    NA
 13          -0.87                  1.00
 14           0.87                  1.00
 15           0.87                 -0.73
 16          -0.87                 -0.73
 17          -0.87                  1.00
 18  Original Curve:
 19     240  -0.87                         -0.50
 20     245  -0.73                         -0.57
  :
 31     300   0.87                         -0.50
 32            NA                            NA
 33       0   0.87                         -0.50
  :
 45      60   0.00                          1.00
 46            NA                            NA
 47          -0.87                         -0.50
  :
 59          -0.00                          1.00
 60  Moving Curve:
 61          -0.87                         -0.50  -0.87  -0.50
 62          -0.73                         -0.57  -0.73  -0.57
  :
101          -0.00                          1.00  -0.00   1.00

G3:  @SQRT(3)                    G4:  @PI/180
C4:  @IF(C3=0,0,5+C4)             C5:  +C4*G4
E3:  @MAX(H61..H101)-G$3/2        E4:  @MAX(I61..I101)-1
B19: 240                          B20: 5+B19
C19: +G$3*@COS(B19*G$4)           F19: 1+G$3*@SIN(B19*G$4)
C33: -G$3/2+G$3*@COS(B33*G$4)     F33: -0.5+G$3*@SIN(B33*G$4)
B33:  0    B34: 5+B33             C47: -C33    F47: +F33
C61: +H61-E$3   H61: @COS(C$5)*C19-@SIN(C$5)*F19
G61: +I61-E$4   I61: @SIN(C$5)*C19+@COS(C$5)*F19
```

Figure 4.70. Curve of Constant Width.

Copy:		Graph:			
From	To	Series	Cells	Labels	Purpose
B20	B21..B31	X	C8..C101		x-axis
C19..F19	C20..F31	1	D8..D11		triangle
B34	B35..B45	2	E8..E17		square
C33..F33	C34..F45	3	F8..F59		curve
C47..F47	C49..F59	4	G8..G101		rotated curve
C61..I61	C62..I101				

Cell C3 is used to initialize the model. When Cell C3 is set to 0, the base curve is created in its original position. When C3 is set to a nonzero value, and with each subsequent recalculation, the curve rotates through a chosen angular increment. In Cell C4 this angle of rotation α is incremented by 5° with each recalculation, and the angle α is converted to radians in Cell C5. The constant $\sqrt{3}$ is generated in Cell G3. The coordinates of the equilateral triangle of sides $\sqrt{3}$ are entered in Rows 8..11 to create the triangle. Here, the points are (0,1), ($-\sqrt{3}/2,-1/2$), and ($\sqrt{3}/2,-1/2$). The square of side $\sqrt{3}$ is created similarly in Rows 13..17. The initial curve is generated in three sections in Rows 19..59. For example, Rows 19..31 generate the arc at the bottom of the triangle, with its center at (0,1), a radius of $\sqrt{3}/2$, and $240° \leq \theta \leq 300°$. The other arcs are created similarly in Rows 33..45 and 47..59, with @NA serving to separate the sides.

Finally, the formulas in Rows 61..101 continually rotate and translate the base curve. First, in Columns H..I each point is rotated through the angle α using the rotation transformation

$$(x,y) \rightarrow (x\cos(\alpha)-y\sin(\alpha), x\sin(\alpha)+y\cos(\alpha))$$

It is important to observe that after entering the formulas in Cells H61,I61 and copying, the entries in those cells in Column H..I that contain @NA from the curve separators must be erased (in Rows 74 and 88) so that @MAX functions that refer to the range do not return @NA.

In order for the curve to stay within the square, it also must be translated. Cells E3 and E4 find the differences A_x and A_y between the maximum current x- and y-values of the curve from Columns H..I and the maximum x and y values of the square. In Columns C and G of Rows 66..101 these differences are subtracted from the rotated values. Thus, as it is rotated and translated, the curve always fits inside of a fixed square, always touching the square's sides.

Activities

1. Modify the model so that the length of the equilateral triangle's side is a parameter.
2. Create other curves of constant width. Several of these are given in Yaglom and Boltyanskii (1961); Cundy and Rollett (1961), pp. 210 to 213; and Bolt (1991), p. 109. In particular, create various n-gon equivalents of Reuleaux triangles.
3. Design an animated model in which the curve of constant width in this section rolls along a line.

4. A Δ-*curve* is a one that can be turned continuously inside an equilateral triangle. Create a model for these curves using the curve bounded by the arcs of circles in Figure 4.71. Information on Δ-curves is found Yaglom and Boltyanskii (1961), pp. 83 to 99, and Honsberger (1973), pp. 54 to 63.

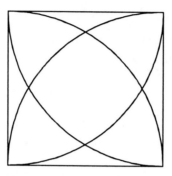

Figure 4.71. Delta Curve.

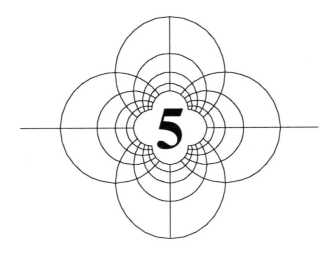

Chapter 5

CURVES FROM COMPLEX VALUED FUNCTIONS

This chapter presents a different approach for creating mathematical curves through the implementation of spreadsheet models of functions of a complex variable. The techniques presented not only provide ways of creating interactive graphical representations for these functions, but they also generate a wide variety of interesting curves. The first two sections discuss properties of complex numbers and aspects of the elementary geometry of complex arithmetic. Sections 5.3 and 5.44 show how to create depictions of complex functions through the use of curves in the plane, while Section 5.5 presents models of the transformation of planar grids. The chapter concludes with some visual applications to fluid flow, airfoils, and similar topics, as well as another approach to using the spreadsheet solver command. The background information on complex variables that is used in this chapter can be found in most introductory texts in the area, including Fisher (1990), Greenleaf (1972), Saff and Snider (1976), Spiegel (1964), and Churchill (1960).

5.1. COMPLEX POWERS

A complex number has the form $z = a+bi$, with a and b real numbers and $i^2 = -1$. A complex number can be represented geometrically in the plane by the point (a,b) as illustrated in Figure 5.1.

The operations of complex *addition* and *multiplication* are defined via the standard algebra definitions, with i^2 replaced by -1. Thus,

$$(a+bi) + (c+di) = (c+b) + (b+d)i \text{ and } (a+bi)(c+di) = (ac-bd) + (ad+bc)i$$

In addition, the *conjugate* of the complex number $a+bi$ is defined as $a-bi$. Complex *division* is carried out using the conjugate of the denominator,

$$\frac{c+di}{a+bi} = \frac{c+di}{a+bi}\frac{a-bi}{a-bi} = \frac{ac+bd}{a^2+b^2} + \frac{ad-bc}{a^2+b^2}i$$

156 Complex Powers

From this it follows that if $z = a+bi$, then $1/z = a/(a^2+b^2) - ib/(a^2+b^2)$.

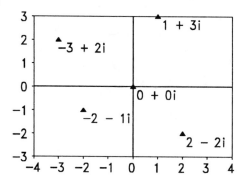

Figure 5.1. Complex Plane.

A complex number $z = a+bi$ can be written in polar form as $z = re^{i\theta} = r\cos(\theta) + ir\sin(\theta)$. The value r is called the *modulus* of z. It represents the distance from the origin to z. Thus,

$$r = |z| = \sqrt{a^2 + b^2}$$

The value θ is called the *argument* of z. It represents the angle that a vector from the origin to the point z makes with the positive x-axis, and is determined by $\tan(\theta) = y/x$. In most spreadsheets this can be determined by @ATAN2(x,y) or by @ATAN(y/x).

Positive integer powers of a complex number z can be created by first forming $z^1 = z$, and then successively computing $z^{n+1} = z(z^n)$ for $n > 0$. Negative integer powers are created similarly by setting $z^{-1} = 1/z$ and then successively calculating $z^{-n-1} = z^{-1}(z^{-n})$ for $n > 0$. Finally, $z^0 = 1+0i$. These definitions are implemented in the model of Figure 5.2.

	A	B	C	D	E	F
1		x	y		r	θ
2	z =	0.650	0.850		1.070	0.918
3	1/z =	0.568	-0.742		0.935	-0.918
4	n	Re	Im	Cir y2	rn	θn
5	-4	-0.658	0.386		0.763	2.611
6	-3	-0.756	-0.309		0.816	-2.754
7	-2	-0.229	-0.843		0.873	-1.836
8	-1	0.568	-0.742		0.935	-0.918
9	0	1.000	0.000		1.000	0.000
10	1	0.650	0.850		1.070	0.918
11	2	-0.300	1.105		1.145	1.836
12	3	-1.134	0.463		1.225	2.754
⋮						
17	8	0.840	1.500		1.719	1.060
18	0	1.000		0.000		
19	1	0.999		0.044		
⋮						
162	144	1.000		-0.000		

```
E2:  @SQRT(B2^2+C2^2)         F2:  @ATAN2(B2,C2)
B3:  +B2/(B2^2+C2^2)          C3:  -C2/(B2^2+C2^2)
A8:  -1+A9    A9:  0    A10:  1+A9    B9:  1    C9:  0
B8:  +B$3*B9-C$3*C9           C8:  +B$3*C9+C$3*B9
B10: +B$2*B9-C$2*C9           C10: +B$2*C9+C$2*B9
A18: 0                        A19: 1+A18
B18: @COS(@PI*A18/72)         D18: @SIN(@PI*A18/72)
```

Figure 5.2. Complex Powers.

Copy:		Graph:			
From	To	Series	Cells	Labels	Purpose
A8..C8	A5..C7	X	B5..B162		x-axis
A10..C10	A11..C17	1	C5..C17	A5..A17	powers
B18..D18	B19..D162	2	D5..D162		circle
E2..F2	E3..F17				
A19	A20..A162				

The model computes z^n for the integers $n = -4,-3,\ldots,8$. First, the coordinates of z are entered in Cells B2..C2. Using the definitions above, the values of r and θ are computed in Cells E2..F2, and the coordinates of $1/z$ are determined in Cells B3..C3. A table of powers is created in Rows 5..17, with Column A used to count the powers. First, $1+0i$ is entered in Cells B9..C9 to form z^0. Next, z^1 is calculated in Cells B10..C10 as $z(z^0)$. Through the use of the $ convention shown, this expression is interpreted as $z(z^{n-1})$ when it is copied down through Row 17 to form the powers z^n. In Cells B8..C8 z^{-1} is computed similarly by using z^{-1} instead of z^1, and then copied up through Row 5 to form the powers z^{-n}. By using the insert and copy technique, the range of powers can be expanded readily. Finally, the formulas in Cells E2..F2 are copied down their columns to calculate the modulus and argument of each power.

The powers of $z = 0.65 + 0.85i$ are illustrated in the graph of Figure 5.3, using Columns B and C as the respective x- and y-series. The unit circle is constructed as an auxiliary curve in Columns B and D of Rows 18..162 using a condensed form of the standard circle techniques of Chapter 2.

By changing the value of z, several facts about complex powers become apparent. For example, the argument of each power is increased from that of the previous one by θ, while the modulus is successively multiplied by r. Thus, with a complex number z whose modulus is greater than 1, each positive power is further from the origin (Figure 5.3), while with $|z| < 1$, the corresponding powers progress inward toward the origin (Figure 5.4). When $|z| = 1$, z and its powers lie on the unit circle. In Figure 5.5 a regular octagon is formed by using $r = 1$ and $\theta = 45°$ ($z = e^{i\pi/4}$). When other angles are used, different patterns result.

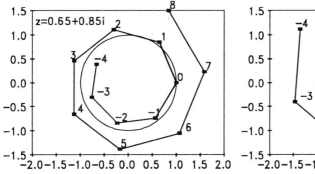

Figure 5.3. Complex Powers, $r > 1$.

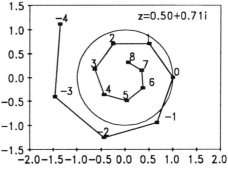

Figure 5.4. Complex Powers, $r < 1$.

158 Complex Powers

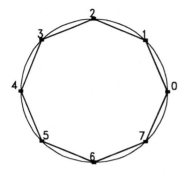
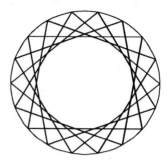

Figure 5.5. Complex Powers: $\theta = 45°$, $r = 1$.   **Figure 5.6.** Complex Powers, $\theta = 100°$, $r = 1$.

Activities

1. Create graphs to depict the sum, difference, product, and quotient of complex numbers.
2. Using Figure 5.2, experiment with points z that lie on or off the unit circle. Use a z with $r = 1$ and $\theta = 100°$ together with the insert and copy commands to create Figure 5.6.
3. Compute powers of a complex number z by $z^n = r^n e^{in\theta} = r^n\cos(n\theta) + ir^n\sin(n\theta)$.
4. Create a model that plots successive partial sums $u_0, u_1, u_2,\ldots$ of a series $\Sigma a_k z^k$, where $u_n = a_0 + a_1 z^1 + \ldots + a_k z^k + \ldots + a_n z^n$. Design a graph showing the convergence or divergence of the partial sums. Figure 5.7 shows one model with $a_k = 1/k$ from which Figures 5.8 and 5.9 were created. Examine: (a) $a_k = k$; (b) $a_k = 1$. Interesting patterns are produced by $z = x+iy$ with $|z| = 1\pm\varepsilon$. In (a) enter $y = \text{sqrt}(1-x^2)+0.0005$ and $x = 0.02$, $0.025, 0.03,\ldots$. In (b) try $x = 0.6\pm\varepsilon$, $y = 0.8$; $x = 0.707\pm\varepsilon$, $y = 0.707$; $x = 0$, $y = 1\pm\varepsilon$.

```
        A      B       C        D        E
1    z=   -0.6    0.9
2     n  Re(z^n) Im(z^n) Re(un)  Im(un)
3     0   1.00    0.00    0.00    0.00
4     1  -0.60    0.90   -0.60    0.90
5     2  -0.45   -1.08   -0.83    0.36
6     3   1.24    0.24   -0.41    0.44

D4: +D3+B4/A4       E4: +E3+C4/A4
```

Figure 5.7. Partial Sums, $a_k = z^k/k$.

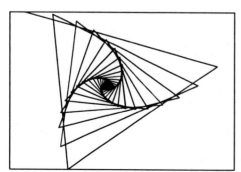
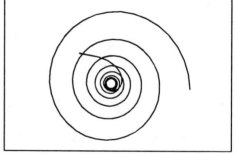

Figure 5.8. Partial Sums, $\Sigma z^k/k$: $z = -0.6+0.9i$.   **Figure 5.9.** Partial Sums, $\Sigma z^k/k$: $z = 1.0-0.19i$.

5.2. COMPLEX ROOTS

If $z = x+yi$ is a complex number and n is a positive integer, then an *n-th root* of z is a complex number w for which $w^n = z$. To find such a root, suppose that both z and w are expressed in polar form. If $w = r_1 e^{\alpha i}$ and $z = re^{\theta i}$, then $w^n = r_1^n e^{n\alpha i} = re^{\theta i}$. Thus, with $r = r_1^n$ and $n\alpha = \theta$, one root is given by

$$w = r_1 e^{\alpha i} = r_1 \cos(\alpha) + ir_1 \sin(\alpha), \text{ where } r_1 = \sqrt[n]{r}, \ \alpha = \Theta/n$$

This particular root w is called a *primitive n-th root* of z. Because, for instance, $\cos(n(\alpha+2\pi k/n)) = \cos(n\alpha+2k\pi) = \cos(\theta)$, there are, in fact, n distinct n-th roots of z that are given by

$$\sqrt[n]{r}\cos(\alpha+2\pi k/n) + i\sqrt[n]{r}\sin(\alpha+2\pi k/n), \quad k = 0,1,\ldots,n-1$$

Each of these complex numbers has the same modulus, while the arguments of the resulting roots differ by $2\pi/n$ for successive values of k. As a consequence, these complex numbers lie on a circle of radius r_1 and form a regular n-gon. Figure 5.10 shows this with $n = 5$ and $z = -0.45 + 0.20i$.

The graph is produced by the model of Figure 5.11, which computes the n-th roots of a complex number z for a positive integer n. The coordinates of z are entered in Cells B2..C2, with n set in Cell B3. Cells D2..E2 determine values for the polar parameters r and θ as in Section 5.1. The corresponding values of r_1 and $\alpha = \theta/n$ for w are computed in Cells D3..E3 using the equations developed above.

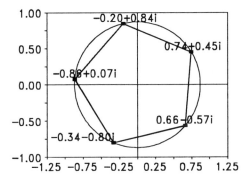

Figure 5.10. 5th Roots of $z = -0.45 + 0.20i$.

A table of roots for $k = 0,1,\ldots,n-1$ is then constructed in Rows 5..17. Column A serves as the counter k. The formula @IF(A5<B$3,1+A5,@NA) is entered in Cell A6 and copied down Column A to produce NA after the counter reaches n. The n-th roots are then found by entering the basic formulas for their coordinates in Cells B5..C5 and copying them down their columns. Column B provides the x-axis for the graph, and Column C the first y-series for the roots. Each of the roots lies on the circle of radius r_1 and center at the origin. This circle is plotted using Rows 18..162 of Columns B and D to generate the graph of Figure 5.10. The interior labels for the roots can be obtained by copying the formula @STRING(B5,2)&@IF(C5<0,"-","+")&@STRING(@ABS(C5),2)&"i" from Cell F5.

160 *Complex Roots*

```
          A          B          C          D        E
  1                  x          y          r        θ
  2     z =       -0.45        0.2       0.492    2.723
  3     n =           5      root:       0.868    0.545
  4       k        Re         Im       Cir y2
  5       0       0.742      0.450
  6       1      -0.198      0.845
  7       2      -0.865      0.073
  8       3      -0.336     -0.800
  9       4       0.657     -0.567
 10       5       0.742      0.450
  :
 17              NA          NA          NA
 18       0       0.868                  0.000
 19       1       0.867                  0.038
  :
162     144       0.868                 -0.000

D2: @SQRT(B2^2+C2^2)       E2: @ATAN2(B2,C2)
D3: +D2^(1/B3)             E3: +E2/B3
A5: 0     A6: @IF(A5<B$3,1+A5,@NA)
B5: +D$3*@COS(E$3+2*@PI*A5/B$3)
C5: +D$3*@SIN(E$3+2*@PI*A5/B$3)
A18: 0                     A19: 1+A18
B18: +D$3*@COS(@PI*A18/72)
D18: +D$3*@SIN(@PI*A18/72)
```

Figure 5.11. Complex Roots.

Copy:		Graph:			
From	To	Series	Cells	Labels	Purpose
A6	A7..A17	X	B5..B162		x-axis
B5..C5	B6..C17	1	C5..C17	F5..F17	roots
A19	A20..A162	2	D5..D162		unit circle
B18..D18	B19..D162				

Activities

1. Amplify the model of Figure 5.11 to allow a user to select an *n*-th root *w*. The model then computes and plots the *n* powers $w^1, w^2, \ldots, w^n$, that progress from $z = w^1$ to $z = w^n$ as shown in Figure 5.12 with a fourth root, $w = 0.74+0.45i$, of $z = -0.45 + 0.20i$.

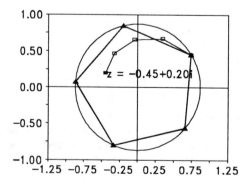

Figure 5.12. Powers of an *n*-th Root.

5.3. GRAPHING COMPLEX FUNCTIONS

A complex function has the form $w = f(z)$, where z and w are complex numbers. Creating a graph for such a function requires some ingenuity, as two dimensions are needed for both the domain and the range. This section describes a model that implements one method for representing a complex function. First, a curve C that bounds an area in the plane is created. Next, from the points z on C, the corresponding $w = f(z)$ values form another curve. In addition to illustrating the effect of the complex function, this process can be used to generate a wide assortment of curves. Figure 5.13 provides an illustrative graph using the function $f(z) = e^z$. The input curve C is the rectangle $1.8 \leq x \leq 3.3$, $-2.5 \leq y \leq \pi$. The resulting image curve is the partial annulus shown.

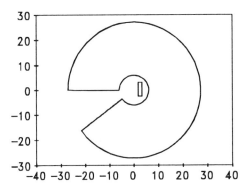

Figure 5.13. $f(z) = e^z$ (via Rectangle).

The model of Figure 5.14 creates the graph. The minimum and maximum x- and y-values for the corners of the base rectangle are entered in Cells F1..H2. The number of points, n, that are computed on each side of the rectangle is set in Cell B1. In this example, $n = 50$. The distances dx and dy between successive x- and y-values on alternate sides of the rectangle are calculated in Cells F3 and H3 by $dx = (x_{hi}-x_{lo})/n$ and $dy = (y_{hi}-y_{lo})/n$. The base input curve (z-values) is generated in Columns E..F of Rows 6..206. The corresponding output curve ($w = f(z)$-values) is created in Columns E and I of Rows 208..408.

Although the four sides of the rectangle can be created separately, the model's formulas generate all of them simultaneously. The multiple curve technique of earlier sections is employed. Four sets of point counters for the range $0,1,...,n-1$ are formed in Column A by entering 0 in Cell A6 and @MOD(1+A6,B$1) in Cell A7. The latter expression is then copied down Column A. The formulas in Column B generate n copies of each of the integers 0, 1, 2, 3 to identify the four sides of the rectangle (0: left, 1: top, 2: right, 3: bottom). This is established by first entering 0 in Cell B6. Then the indicated formula is entered in Cell B7 and copied down Column B.

The coordinates of points on the base rectangle are created in Column E..F. The @CHOOSE function for x in Column E selects the minimum (F$1) or maximum (F$2) x-value when the entry in Column B is 0 or 2, respectively. Otherwise, it increments the previous x-value by the amount dx (F$3). The increment is positive when the Column B entry is 1, and negative if it is 3. The formulas for y in Column F are interpreted similarly. The graph of the rectangle in Figure 5.13 is created with Column E providing x-values and Column F giving the first y-series. Finally, the values for the polar coordinates r and θ of each point z are found in Columns G..H using the standard definitions.

162 Graphing Complex Functions

Next, the image curve is created in Rows 208..408 of Columns E and I by calculating values by $w = f(z) = u+iv$ for each point z in C. Because

$$w = e^z = e^{x+iy} = e^x\cos(y) + ie^x\sin(y)$$

the value of u for the first point $w = f(z)$ is computed as @EXP(E6)*@COS(F6) in Cell E208, while the initial v is determined in Cell I208 as @EXP(E6)*@SIN(F6). These expressions are then copied down their respective columns. To form the output curve, the x-series is extended down Column E, with Column I used as the second y-series. The input and output curves are displayed simultaneously in Figure 5.13. After the model has been created, changing the rectangle's parameters alters the resulting figure. For example, an entire annulus (i.e., washer) is produced by using $-\pi \le y \le \pi$ (Cells H2..H3).

```
         A    B   C  D   E       F       G       H       I
   1 n =  50              lo x=   1.8  lo y=  -2.5
   2                      hi x=   3.3  hi y=   3.14
   3                      dx  =   0.03 dy  =   0.11
   4 Domain:                   Input                      Output
   5    n    m  θ1 r1     x       y1      r       θ       y2
   6    0    0             1.80  -2.50  3.08  -0.95
   7    1    0             1.80  -2.39  2.99  -0.92
   :
  55   49    0             1.80   3.03  3.52   1.03
  56    0    1             1.83   3.14  3.64   1.04
  57    1    1             1.86   3.14  3.65   1.04
   :
 155   49    2             3.30  -2.50  4.14  -0.65
 156    0    3             3.27  -2.50  4.12  -0.65
 157    1    3             3.24  -2.50  4.09  -0.66
   :
 205   49    3             1.80  -2.50  3.08  -0.95
 206    0    4             ERR    ERR   ERR    ERR
 207 Image:
 208    0    0            -4.85                         -3.62
 209    1    0            -4.41                         -4.14
   :
 407   49    3            -4.85                         -3.62

F3:  (F2-F1)/B1    H3: (H2-H1)/B1
A6:  0      A7: @MOD(1+A6,B$1)
B6:  0      B7: +B6+(A7=0)
E6:  +F1    E7: @CHOOSE(B7,F$1,E6+F$3,F$2,E6-F$3)
F6:  +H1    F7: @CHOOSE(B7,F6+H$3,H$2,F6-H$3,H$1)
G6:  @SQRT(E6*E6+F6*F6)          H6: @ATAN2(E6,F6)
A208: +A6   B208: +B6    E208: @EXP(E6)*@COS(F6)
I208: @EXP(E6)*@SIN(F6)
```

Figure 5.14. Exponential Function on Rectangle.

Copy:		Graph:			
From	To	Series	Cells	Labels	Purpose
A7..F7	A8..F206	X	E6..E408		x-axis
G6..H6	G7..H206	1	F6..F206		input curve
A208..I208	A209..I408	2	I6..I408		output curve

It is easy to use other figures as the base input curve. Making the changes indicated in Figure 5.15 using steps of 2.5° produces a circle of radius r and center (h,k) as the input curve. Various interesting curves, such as Figure 5.16, are produced simply by changing the model's parameters in Cells D1..D3. The model of Figure 5.15 reserves Columns C and D for the use of the polar parameters θ_1 and r_1 to describe a curve when they are needed.

Among the other possible input contours are triangles, sectors, strips, and half-planes, as well as any of the curves that are presented in Chapter 2. Figure 5.17 uses a variable triangle for C. In addition, the model can be adapted for other functions by simply changing the formulas in Row 152 and copying them appropriately.

```
         A       B        C       D       E       F       G       H       I
    1 Circle:            r =      4
    2 α = 0.044          h =     -2
    3                    k =      0
    4 Domain:
    5    n               θ1      r1      x       y1      r       θ       y2
    6    0                0              2.00    0.00    2.00    0.00
    7    1               0.044           2.00    0.17    2.00    0.09
    8    2               0.087           1.98    0.35    2.02    0.17
    :
  150  144               6.283           2.00   -0.00    2.00   -0.00
  151 Image:
  152    0                               7.39                                   0.00
  153    1                               7.25                                   1.28
  154    2                               6.84                                   2.49
    :
  296  144                               7.39                                  -0.00

  B2:   2.5*@PI/180                  A6:   0    A7:  1+A6
  C6:   +A6*B$2                      E6:   +D$2+D$1*@COS(C6)
  F6:   +D$3+D$1*@SIN(C6)            G6:   @SQRT(E6*E6+F6*F6)
  H6:   @ATAN2(E6,F6)                A152: +A6
  E152: @EXP(E6)*@COS(F6)            I152: @EXP(E6)*@SIN(F6)
```

Figure 5.15. Exponential Function $f(z) = e^z$ (via Circle).

Copy:		Graph:			
From	To	Series	Cells	Labels	Purpose
A7	A8..A150	X	E6..E296		x-axis
C6..H6	C7..H150	1	F6..F150		input curve
A152..I152	A153..I296	2	I6..I296		output curve

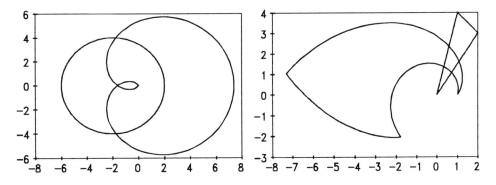

Figure 5.16. $f(z) = e^z$, Base: Circle.

Figure 5.17. $f(z) = e^z$, Base: Triangle.

Activities

1. Experiment with the parameters of the two models of this section to generate various curves. In addition, design and implement alternate curves and shapes for the base curve C. For example, use curves discussed in Chapter 2, or a line, for C.
2. Create models for other functions $w = f(z)$. Figure 5.18 uses a rectangle with the function $f(z) = z^2 + z$. Figure 5.19 uses a rectangle not containing the origin and the square root function $f(z) = z^{1/2}$. Its values are computed by using polar coordinates to express z as $re^{i\theta}$, and determining $f(z)$ as $r^{1/2}e^{i\theta/2}$.

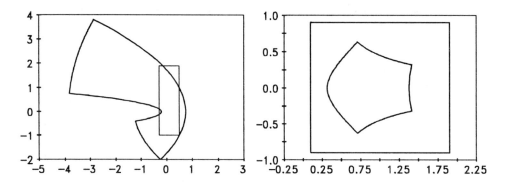

Figure 5.18. $f(z) = z^2 + z$.   **Figure 5.19.** $f(z) = z^{1/2}$.

5.4. ADDITIONAL COMPLEX FUNCTIONS

This section displays graphs created from several other complex functions. These graphs can be created through slight modifications of the previous models. The graph of Figure 5.20 illustrates the effect of the function $f(z) = z^2$ on a sector of a circle whose center lies at the origin.

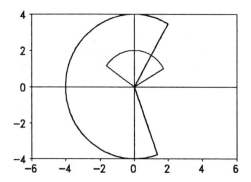

Figure 5.20. $f(z) = z^2$, Base: Sector.

The model of Figure 5.21 produces the graph above by creating a sector of a circle as the base curve in Rows 6..250. The sector is built in three parts using the parameters and values in the top three rows. The sector's arc is formed in Rows 6..150 by using the standard technique for drawing a circle. The radius r of the sector is entered in Cell B3. The initial and terminal angles θ_1 and θ_2 in degrees are entered in Cells B1..B2, and then changed to radians in Cells D1..D2. Because 144 points are used to plot the arc, the step size between successive values of θ is computed in Cell D3 as $d\theta = (\theta_2-\theta_1)/144$. Thus, in Rows 6..150 of Column C, values of θ are calculated by starting at θ_1 and successively adding $d\theta$, so that θ ranges from θ_1 to θ_2. The x- and y-coordinates of the arc are found in Rows 6..150 of Columns E..F as $r\cos(\theta)$ and $r\sin(\theta)$.

```
           A     B       C      D        E       F        G
     1   θ1=    30     θ1= 0.524                SECTOR
     2   θ2=   145     θ2= 2.531                W=Z^2
     3   r =     2     dθ= 0.014
     4   Domain:
     5     n             θ                 x       y1       y2
     6     0    arc    0.52              1.73    1.00
     7     1           0.54              1.72    1.02
     8     2           0.55              1.70    1.05
     :
   150   144    end    2.53             -1.64    1.15
   151     1 right                      -1.61    1.12
   152     2                             -1.57    1.10
     :
   200    50    end                     -0.00   -0.00
   201     1 left                        0.03    0.02
   202     2                             0.07    0.04
     :
   250    50    end                      1.73    1.00
   251   Image:                            NA      NA
   252     0                             2.00            3.46
   253     1                             1.90            3.52
     :
   496    50                             2.00            3.46

   D1:   +B1*@PI/180           D2:  +B2*@PI/180
   D3:   (D2-D1)/144
   A6:     0    A7:  1+A6      C6:  +D1    C7:  +C6+D$3
   E6:   +B$3*@COS(C6)         F6:  +B$3*@SIN(C6)
   E151: +E150-E$150/50        F151: +F150-F$150/50
   E201: +E200+E$6/50          F201: +F200+F$6/50
   E252: +E6^2-F6^2            G252: 2*E6*F6
   A151: +A7   A201: +A7       A252: +A6
```

Figure 5.21. $f(z) = z^2$, Base: Sector.

Copy:		Graph:			
From	To	Series	Cells	Labels	Purpose
A7..C7	A8..C150	X	E6..E496		x-axis
E6..F6	E7..F150	1	F6..F250		domain
A151..F151	A152..F200	2	G6..G496		range
A201..F201	A202..F250				
A252..G252	A253..G496				

The sector's bounding radii are created using 50 points each by entering the indicated formulas in Rows 151 and 201 and copying them appropriately. For example, one of the arc's endpoints, (X,Y), appears in Cells E150..F150. By dividing the radius line into 50 points, the x- and y-coordinates are incremented by $X/50$ and $Y/50$, respectively. Thus, successive points (x_i, y_i), $i = 0, 1, \ldots, n$, on the radius can be calculated by

$$x_0 = X,\ y_0 = Y,\ x_{n+1} = x_n - X/50,\ y_{n+1} = y_n - Y/50.$$

A similar process repeated in Row 201 for the other bounding radius line.

Finally, for each z in Rows 6..250 the coordinates of the points of $w = f(z) = z^2$ are calculated in Rows 252..496 of Columns E and G by using $z^2 = (x+yi)^2 = (x^2-y^2) + 2xyi$. The graph is created by using Column E as the x-axis series, and Columns F and G for the y-values of the base circle and the resulting curve, respectively.

By using this graph while varying the parameters r, θ_1, and θ_2, it is easy to observe certain features of the function $f(z) = z^2$. For example, the distance from origin to the point $f(z)$ is the square of the distance from the origin to z, and the corresponding angles made by lines from the origin to the points are doubled. It is straightforward to adapt the model for functions $f(z) = z^n$ for other n, for example, $f(z) = z^3$.

A second useful way to study the function $f(z) = z^n$ is by examining its action on a circle. This is also a particularly effective way to produce curves. Figure 5.22 shows a model for doing this with $f(z) = z^3$. If $z = a+bi$, then $z^3 = (a^3 - 3ab^2) + (3a^2b - b^3)i$. This model produces Figure 5.23. Figure 5.24 is formed from a similar model using $f(z) = z^2$. By varying the placement of the base circle, a great variety of intriguing curves can be produced.

```
         A        B        C        D       E
    1  r  =     1.5     h  =    -0.7    CIRCLE
    2  α  =   0.022     k  =     0.0    W = Z³
    3  Domain:
    4    n        θ            x        y1        y2
    5    0        0         0.80     0.00
    6    1    0.022         0.80     0.03
    :
  293  288    6.283         0.80    -0.00
  294  Image:
  295    0                  0.51              0.00
  296    1                  0.51              0.06
    :
  583  288                  0.51             -0.00

  B2: 1.25*@PI/180         B5: +A5*B$2
  A5: 0                    A6: 1+A5
  C5: +D$1+B$1*@COS(B5)
  D5: +D$2+B$1*@SIN(B5)
  A295: +A5   C295: +C5^3-3*C5*D5^2
  E295: 3*C5^2*D5-D5^3
```

Figure 5.22. $f(z) = z^3$, Base: Circle.

Copy:		Graph:			
From	To	Series	Cells	Labels	Purpose
A6	A7..A293	X	C5..C583		x-axis
B5..D5	B6..D293	1	D5..D293		input curve
A295..E295	A296..E583	2	E5..E583		output curve

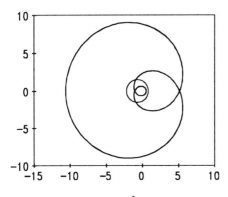
Figure 5.23. $f(z) = z^3$, Base: Circle.

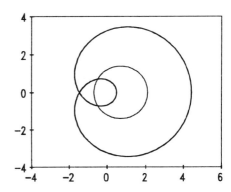
Figure 5.24. $f(z) = z^2$, Base: Circle.

Another function that generates informative curves is $f(z) = 1/z$. It is shown in the study of complex analysis that this function always maps circles into circles or lines. If the base circle in the domain lies inside the unit circle $x^2 + y^2 = 1$, then its image lies outside of it, and conversely. If the base circle passes through the origin then the image is a line. It is instructive to compare the curves here with those produced in Section 3.6. In Figures 5.25 through 5.28 the unit circle is C, the domain circle is A, and the image curve is B.

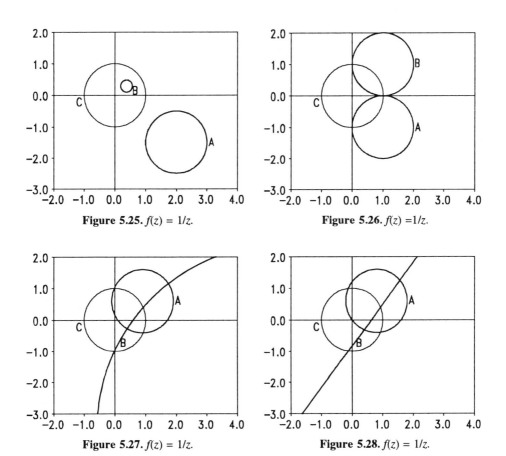

Figure 5.25. $f(z) = 1/z$.

Figure 5.26. $f(z) = 1/z$.

Figure 5.27. $f(z) = 1/z$.

Figure 5.28. $f(z) = 1/z$.

168 *Additional Complex Functions*

Finally, a number of fundamental geometric transformations can be produced through functions of the forms $f(z) = z + \beta$, $f(z) = \alpha z$, or $f(z) = \alpha z + \beta$, where α and β are complex numbers. The previous models can be adapted to these functions. However, by using simple figures that are bounded by straight line segments for the base curve, a much more compact model can be designed. The effect of the function $f(z) = z + a$, with $a = 1+2i$, on a rectangle is shown in Figure 5.29. This graph was produced by the model of Figure 5.31. It shows that the function translates the rectangle *abcd* by 1 unit in the *x*-direction and 2 units in the *y*-direction to *ABCD*. The base rectangle is created via formulas. Figure 5.30 shows how more complex objects can be used, this time via a rotation and a translation.

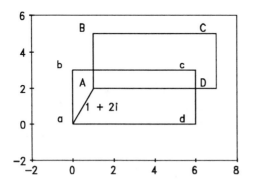
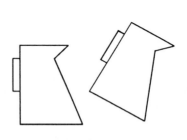

Figure 5.29. Translation. **Figure 5.30.** $f(z) = \alpha z + \beta$.

```
              A         B         C         D
 1  Pt: a  =           1    b  =           2
 2  Rc:x0  =           0    y0 =           0
 3       dx =          6    dy =           3
 4  Graph:
 5          x         y1        y2  Labels
 6          0          0                  a
 7          0          3                  b
 8          6          3                  c
 9          6          0                  d
10          0          0
11          1                    2  A
12          1                    5  B
13          7                    5  C
14          7                    2  D
15          1                    2

A11: +A6+B$1    C11: +B6+D$1
```

Figure 5.31. Translation.

Copy:		Graph:			
From	To	Series	Cells	Labels	Purpose
A11..C11	A12..C15	X	A6..A15		x-axis
		1	B6..B10	D6..D9	input rectangle
		2	C6..C15	D6..D14	output rectangle

Activities

1. Implement each of the models in this section, experimenting with various base curves.
2. Create models for (a) rotations, $f(z) = \alpha z$, with $|\alpha| = 1$; (b) dilations, $f(z) = az$, with $0 < a < 1$; (c) other transformations of the form $f(z) = \alpha z + \beta$, where α and β are complex numbers; (d) polynomials, $f(z) = a_n z^n + \ldots + a_2 z^2 + a_1 z + a_0$; (e) trigonometric functions such as $f(z) = \sin(z)$, where $\sin(x+iy) = \sin(x)(e^y + e^{-y})/2 + i\cos(x)(e^y - e^{-y})/2$.
3. Create a model for the linear fractional transformations $f(z) = (\alpha z + \beta)/(\sigma z + \mu)$, where the constants are complex numbers. Study the effects of the resulting transformations on lines and circles in particular.
4. Design a more complex base diagram such as in Figure 5.30.
5. Just as with functions of a complex variable, linear transformations from linear algebra and affine transformations of $\mathbf{R}^2$ can be used to transform curves. Create a model to investigate the effects of such linear transformations as reflections, dilations, shears, and others on different base curves. Educational aspects of this topic are found in Carter (1992).

5.5. GRID REPRESENTATIONS OF FUNCTIONS

Another way to visualize a complex function is by showing its effect upon the lines that make up a coordinate grid system for the *z*-plane. The model of Figure 5.34 does this using the standard rectangular axes. The model creates thirteen horizontal lines and thirteen vertical lines to provide a grid for a rectangular region of the complex plane, as in Figure 5.32. The model then evaluates the exponential function, $f(z) = e^z$, at the points on these grid lines to produce the curves shown in Figure 5.33. Because of the large number of points that must be plotted, the resulting spreadsheet model is rather sizable.

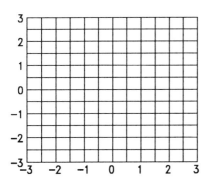

Figure 5.32. Domain, $f(z) = e^z$.

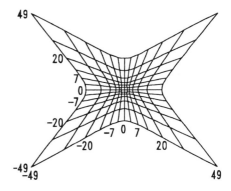

Figure 5.33. Range, $f(z) = e^z$.

To use the model, x_0 and y_0, the minimum values for x and y, are entered in Cells D3 and F3. Here, $x_0 = y_0 = -3$. Next, h_x and h_y, the distances between the x- and y-grid lines of the domain, are entered in Cells D1..D2. Here, 0.5 is used for each of these to produce the lines

$x = -3.0$, $x = -2.5,\ldots$, $x = 3.0$, and $y = -3.0$, $y = -2.5,\ldots$, $y = 3.0$. The multiple line technique of Section 1.6 is used twice to create the vertical lines in Rows 5..498 and the horizontal lines in Rows 499..992. To see the plan for creating Rows 5..498, let $i = 0,1,2,\ldots,12$ count the lines (Column A) and $j = 0,1,2,\ldots,37$ count points on each line (Column B). Points are plotted along the lines at a frequency of $\alpha_y = 1/3$ (Cell F2) of the y-grid size. The (x,y)-coordinates are set to @NA when the entry in Column B is 37 to produce line separators. Otherwise, if the x-values in Column C are denoted by x_i and the y-values in Column D by y_j, then $x_i = x_0 + ih_x$ and $y_j = y_0 + j\alpha_y h_y$. Thus, the (x,y)-coordinates of the grid are formed by entering the indicated formulas in Cells C5..D5 and copying them through Row 498. The horizontal lines are created similarly in Rows 499..992. Columns C and D define the grid for the function's domain in Figure 5.32.

In this approach, the domain and range are not shown simultaneously. Thus, the model can be designed to use the same rows for the coordinates of both z and $f(z)$. Consequently, the real component u of the image of the first point $u+iv = f(x+iy) = e^x\cos(\theta) + ie^x\sin(\theta)$ is found in Cell E5 as

$$u = \exp(r)\cos(\Theta) = \exp(\sqrt{x^2+y^2})\cos(\operatorname{atan2}(x,y))$$

The corresponding imaginary component v is found similarly in Cell F5. These formulas are then copied down their columns through Row 498. Finally, the image of the domain's grid is produced to give the range by using Columns E and F for the x- and y-series.

```
          A      B      C        D      E          F
    1   EXP    x-grid =         0.5  x-step =   0.333
    2   GRID   y-grid =         0.5  y-step =   0.333
    3                  x0 =      -3   y0 =         -3
    4   x-ct   y-ct    x         y     u            v
    5    0      0      0      -3.00  -3.00      -49.21   -49.21
    6    0      1     -3.00   -2.83  -45.05     -42.54
    7    0      2     -3.00   -2.67  -41.38     -36.78
    :
   41    0     36     -3.00    3.00  -49.21      49.21
   42    0     37      NA       NA     NA          NA
   43    1      0     -2.50   -3.00  -31.79     -38.15
   44    1      1     -2.50   -2.83  -28.95     -32.81
    :
   79    1     36     -2.50    3.00  -31.79      38.15
   80    1     37      NA       NA     NA          NA
   81    2      0     -2.00   -3.00  -20.41     -30.62
    :
  497   12     36      3.00    3.00   49.21      49.21
  498   12     37      NA       NA     NA          NA
  499    0      0     -3.00   -3.00  -49.21     -49.21
  500    1      0     -2.83   -3.00  -42.54     -45.05
  501    2      0     -2.67   -3.00  -36.78     -41.38
    :
  991   36     12      3.00    3.00   49.21      49.21
  992   37     12      NA       NA     NA          NA

A5:   0       A6:  +A5+(B6=0)
B5:   0       B6:  @MOD(1+B5,38)
C5:  @IF(B5=37,@NA,D$3+A5*D$1)
D5:  @IF(B5=37,@NA,F$3+B5*F$2*D$2)
E5:  @EXP(@SQRT(C5^2+D5^2))*@COS(@ATAN2(C5,D5))
F5:  @EXP(@SQRT(C5^2+D5^2))*@SIN(@ATAN2(C5,D5))
A499: 0       A500: @MOD(1+A499,38)
B499: 0       B500: +B499+(A500=0)
C499: @IF(A499=37,@NA,D$3+A499*D$1*F$1)
D499: @IF(A499=37,@NA,F$3+B499*D$2)
```

Figure 5.34. $f(z) = e^z$ (via Grid).

Copy:		Graph:			
From	To	Series	Cells	Labels	Purpose
A6..B6	A7..B498	X	C5..C992		x-axis
A500..B500	A501..B992	1	D5..D992		domain
C5..D5	C6..D498	---	-------		-------
C499..D499	C500..D992	X	E5..E992		x-axis
E5..F5	E6..F992	1	F5..F992		range

Once the model is created, the grid step sizes (Cells D1,D2) and the initial x- and y-values (Cells D3,F3) can be reset to display the effect of the function on different areas of the z-plane. It is also easy to modify the model for other functions. Figures 5.35 and 5.36 illustrate the sine and inverse functions using the same base grid. In Figures 5.37 and 5.38 the base grid is shifted to illustrate that the upper half-plane is mapped onto the unit disk by $f(z) = i(z-i)/(z+i)$. The polar grid illustrated in Section 2.3 also can be recreated in the fashion above to illustrate complex functions. Figures 5.39 and 5.40 employ this grid with the function $f(z) = z^2+1$.

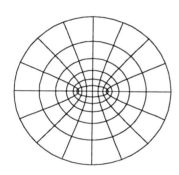

Figure 5.35. $f(z) = \sin(z)$.

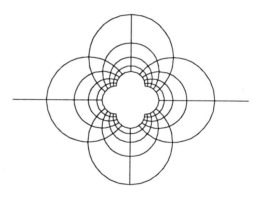

Figure 5.36. $f(z) = 1/z$.

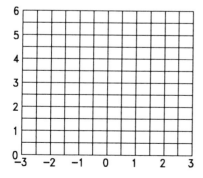

Figure 5.37. Half-Plane.

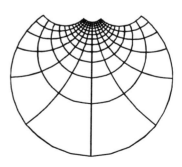

Figure 5.38. $f(z) = i(z-i)/(z+i)$.

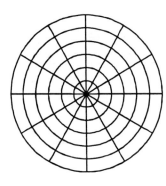

Figure 5.39. Polar Grid.

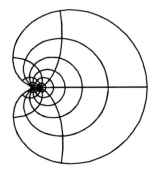

Figure 5.40. $f(z) = z^2+1$.

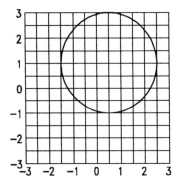

Figure 5.41. Domain Grid and Curve.

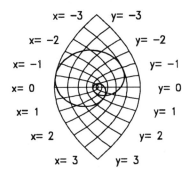

Figure 5.42. $f(z) = z^2+1$.

Activities

1. Create the model of Figure 5.34. Experiment with different grid sizes and functions. Expand the models to include diverse curves shown along with the grid, such as the circle which is used with the function $f(z) = z^2$ in Figures 5.41 and 5.42. For additional examples to pursue, see Greenleaf (1972), pp. 199 to 200, 224, and 239.
2. Use the model of Figure 5.34 to relocate the domain's grid so that it is not centered at the origin.
3. Create a model to illustrate a complex function by using a polar coordinate grid. It is possible to do this by modifying the model of Section 2.3. Use the resulting model with various functions, as in Figures 5.39 and 5.40. Also, try $f(z) = \sin(z)$ and $f(z) = \cos(z)$. In addition consider ways to implement other grids defined by families of orthogonal curves (i.e., $y = c_1/x$ and $x^2-y^2 = c_2$). The latter can be produced from the polar grid and $f(z) = z^2$ (Fisher, 1990, p. 214), and then can be used as the base grid for another function.

5.6. STREAMLINES

A function f is *conformal* at z if for each pair of curves C_1 and C_2 passing through z, the angle between C_1 and C_2 at z is the same as the angle between their image curves C_1' and C_2' at $f(z)$. There are many applications of conformal mappings within the physical sciences. In this section we examine two models related to fluid flow streamlines. For the study of such problems, many books on complex analysis, including those listed at the beginning of the chapter, list conformal functions that map one area of the plane into another.

5.6.A. Flow Over a Constriction

The first model examines water flow over a vertical constriction. (See Saff and Snider, 1976, pp. 350 to 351.) The basic concept consists of mapping the upper half-plane into the upper half-plane so that the x-axis is transformed into the constricted base of the stream. The function then maps lines that are parallel to the x-axis onto the appropriate streamlines. This is illustrated in Figures 5.43 and 5.44. The mapping that accomplishes the task is

$$f(z) = \sqrt{z^2-1}$$

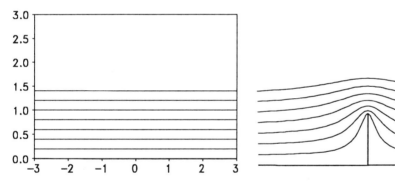

Figure 5.43. Domain Lines.   **Figure 5.44.** Flow Over Constriction.

The domain's graph is created by the model of Figure 5.45 using the multiple line technique of Section 1.6 to produce horizontal lines. Each of these lines contains 61 points, proceeding from -3 to 3 in steps of 0.1. Column A counts the lines, while Column B counts points on the lines. To generate the counters, first 0 is entered in Cells A4 and B4. Then the indicated formulas are entered in Cells A5 and B5 and copied down through Row 498. Values of x are calculated in Column C using steps of 0.1 starting from -3.0. The formulas in Column C create NA when the point counter in Column B is 61, generating a line spacer. A value for the size of the gaps between the lines is entered in Cell D1. By multiplying the current line counter in Column A by D1, y-values for the lines are formed in Column D.

Next, Columns E and F compute the real and imaginary components of the intermediate values $z^2-1 = (x+yi)^2-1 = (x^2-y^2-1)+2xyi$. Denoting these coordinates as (x_1,y_1), the polar coordinates r and θ are determined in Columns G..H as

$$r = \sqrt{x_1^2+y_1^2}, \quad \Theta = \arctan2(x_1,y_1)$$

174   *Streamlines*

Finally, the square root of z^2-1 is computed in Columns I..J for positive values of x using

$$\sqrt{r}\,e^{i\Theta/2} = \sqrt{r}\cos(\Theta/2) + i\sqrt{r}\sin(\Theta/2)$$

Each nonzero complex number has two square roots. To produce continuous curves the formulas in Columns I..J an @IF construction is used to compute the square root using the angles $\pi+\theta/2$ for negative values of x. To avoid continuity problems that arise along the x-axis, in Column D the horizontal lines are translated upwards slightly by using a small offset of 1.0E-15, set in Cell F1. To complete the model, the indicated formulas are entered in Cells C4..J4 and copied down their columns.

The domain's graph is formed by using the points (x,y) of Columns C..D for its x- and the first y-series, while the range's graph uses the points (u,v) of Columns I..J for its x- and first y-series. Cell D1 is used to vary the y-gap between the domain's lines.

```
     A  B   C    D     E    F    G    H     I    J
1        y-step= 0.2  eps = 1E-15
2           Baselines |-auxiliary work-|  Stream's
3    m  n   x    y    x1   y1   r    θ     u    v
4    0  0  -3.0  0.0  8.00 0.00 8.00 0.00 -2.8  0.0
5    0  1  -2.9  0.0  7.41 0.00 7.41 0.00 -2.7  0.0
6    0  2  -2.8  0.0  6.84 0.00 6.84 0.00 -2.6  0.0
:
64   0 60   3.0  0.0  8.00 0.00 8.00 0.00  2.8  0.0
65   0 61   NA   NA    NA   NA   NA   NA   NA   NA
66   1  0  -3.0  0.2  7.96 -1.2 8.05 0.15 -2.8  0.2
67   1  1  -2.9  0.2  7.37 -1.2 7.46 0.16 -2.7  0.2
:
498  7 60   3.0  1.4  6.04  8.4 10.3 0.95  2.9  1.5

A4: 0    A5: +A4+(B5=0)     B4: 0     B5: @MOD(1+B4,62)
C4: @IF(B4=61,@NA,-3+B4/10)           E4: +C4^2-D4^2-1
D4: @IF(B4=61,@NA,A4*D$1+F$1)         F4: 2*C4*D4
G4: @SQRT(E4^2+F4^2)                  H4: @ATAN2(E4,F4)
I4: @SQRT(G4)*@COS(H4/2+@PI*(C4<0))
J4: @SQRT(G4)*@ABS(@SIN(H4/2))
```

Figure 5.45. Fluid Flow Over Constriction.

Copy:		Graph:			
From	To	Series	Cells	Labels	Purpose
A5..B5	A6..B498	X	C4..C498		x-axis
C4..J4	C5..J498	1	D4..D498		domain
		---	--------		--------
		X	I4..I498		x-axis
		1	J4..J498		range

5.6.B. Flow in a Channel

Similar applied routines and physical models can be found in a wide range of books on complex analysis. For example, Figure 5.46 (Saff and Snider, 1976, p. 356) illustrates fluid flow in a blocked channel by using the function $f(z) = \arcsin(z)$ together with the same domain as the prior model.

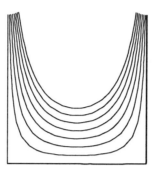

Figure 5.46. Channel Flow.

A brief outline of one possible model is provided in Figure 5.47. The arcsin function is defined by

$$\arcsin(z) = -i\log(iz+\sqrt{1-z^2})$$

Several columns are used in generating intermediate steps to produce the real and imaginary components (u,v) of $\arcsin(z) = u + iv$ using the definition above, with

$$\log(w) = \ln(|w|) + i\arg(w)$$

Columns A and B provide line and point counters, respectively. Columns C and D then generate the (x,y) coordinates of points z of several horizontal lines. Columns E and F give the coordinates (x_1,y_1) of $w = 1-z^2$, Columns G and H calculate the polar coordinates r,θ of w, and Columns I and J provide the coordinates (x_2,y_2) of $w_2 = iz + w^{1/2}$. Finally, Columns K and L provide the coordinates (u,v) of $\arcsin(z)$ by

$$u = \arg(w_2) = \arctan(y_2/x_2), \quad v = -\ln(|w_2|) = -\ln(\sqrt{x_2^2+y_2^2})$$

The formulas contain two technical modifications of this basic scheme. A slight offset (set in Cell F1) is used in Columns G and H to ensure that a point is generated for corners of the channel boundaries. The variant for A4=0 (as in Cell L4) in Column L is needed only for points on the x-axis in the domain. This ensures that the channel boundaries extend in the positive y-direction. The copy and graph commands are straightforward to implement. Columns C..D define the domain's graph, while Columns K..L define the graph of the range. Copy and graph commands are similar to those for Figure 5.45, with Columns K and L providing the streamlines.

Another approach to creating the graph using a spreadsheet solver is described in the activities for Section 5.10.

176   Streamlines

```
      A    B    C      D     E     F       G     H     I     J      K       L
1          y-step =   0.1  eps= 1E-15
2  Count  |-Baselines-|---- Auxiliary computations --|-Streamlines-|
3    m    n    x      y     x1    y1      r     θ     x2    y2     u       v
4    0    0   -3.0   0.0   -8.0   0.00  8.00  3.14   0.0  -0.2   -1.57    1.76
5    0    1   -2.9   0.0   -7.4   0.00  7.41  3.14   0.0  -0.2   -1.57    1.73

A4: 0    A5: +A4+(B5=0)
C4: @IF(B4=61,@NA,-3+B4/10)              B4: 0    B5: @MOD(1+B4,62)
E4: 1-C4^2+D4^2                          D4: @IF(B4=61,@NA,A4*D$1)
G4: @SQRT(E4^2+F4^2+F$1)                 F4: -2*C4*D4
I4: -D4+@SQRT(G4)*@COS(H4/2)             H4: @ATAN2(E4,F4+F$1)
K4: @ATAN2(I4,J4)                        J4: +C4+@SQRT(G4)*@SIN(H4/2)
L4: @IF(A4=0,@ABS(@LN(@SQRT(I4^2+J4^2))),-@LN(@SQRT(I4^2+J4^2)))
```

Figure 5.47. Blocked Channel Flow.

Activities

1. Create the model of Figure 5.45. Design a means of changing the size of the vertical constriction.
2. Create the model of Figure 5.47. Develop similar physical models described in complex variable books: Saff and Snider (1976), pp. 357 to 365; Fisher (1990), pp. 219 to 244; Greenleaf (1972), pp. 503 to 524; Churchill (1960), pp. 209 to 217; and Spiegel (1964), pp. 236 to 237.

5.7. JOUKOWSKI AIRFOIL

One conformal mapping application lies in the design of airfoils. (See Fisher, 1990, pp. 275 to 278; Saff and Snider, 1976, pp. 301 to 302.) The *Joukowski airfoil* of Figure 5.48 is formed via the function $f(z) = az + b/z$ from the circle with center at (x_0, y_0) and radius r, where

$$r = \sqrt{(R-x_0)^2 + y_0^2}$$

and $R^2 = b$. The graph shown employs $a = 1$ and $b = 3$, with the center of the circle located at $(-0.3, 0.4)$. The model of Figure 5.49 creates this airfoil, and allows a user to vary the parameters a, b, x_0, and y_0 to produce a multiplicity of variations.

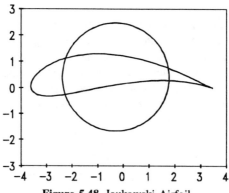

Figure 5.48. Joukowski Airfoil.

To create the model of Figure 5.49, values of parameters are entered in Rows 2..4 as shown, with R calculated in Cell D4 as @SQRT(H4). The radius of the base circle is computed in Cell D5 using the definition above. Points z of the base circle are created in Columns D..E of Rows 7..79 with $x = x_0 + r\cos(\theta)$ and $y = y_0 + r\sin(\theta)$. Using complex arithmetic, the coordinates of $w = f(z) = u + iv$ for these points are first found in Columns G and H as $u = x(a+b/(x^2+y^2))$ and $v = y(a-b/(x^2+y^2))$. These are then reproduced in Columns D and F of Rows 80..152. The appropriate formulas are entered in Row 7 and copied down their columns. The graph of Figure 5.48 is defined with Column D as the x-axis series, and Columns E and F as the y-series for the circle and airfoil, respectively.

```
        A      B      C       D       E       F       G       H
1  JOUKOWSKI  AIRFOIL
2  Center:          x0 =   -0.3            w = az + b/z
3                   y0 =    0.4            a =                1
4  Radius:          R  =    1.73           b = R² =           3
5                   r  =    2.07           pi/180 =        0.02
6   θ:d    θ:r              x       y1              y2      u       v
7    0      0              1.77    0.40             3.38   0.04
8    5    0.09             1.76    0.58             3.30   0.08
:
79  360   6.28             1.77    0.40             3.38   0.04
80    0                    3.38            0.04
81    5                    3.30            0.08
:
152 360                    3.38            0.04

D4: @SQRT(H4)       D5: @SQRT((D4-D2)^2+D3^2)
A7: 0    A8: 5+A7   B7: +A7*H$5    H5: @PI/180
D7:+D$5*@COS(B7)+D$2    E7:+D$5*@SIN(B7)+D$3
G7:+D7*(H$3+H$4/(D7^2+E7^2))
H7:+E7*(H$3-H$4/(D7^2+E7^2))
A80: +A7            D80: +G7                F80: +H7
```

Figure 5.49. Joukowski Airfoil.

Copy:		Graph:			
From	To	Series	Cells	Labels	Purpose
A8	A9..A79	X	D7..D152		x-axis
B7..H7	B8..H79	1	E7..E79		circle
A80..F80	A81..F152	2	F9..F152		airfoil

Activities

1. Experiment with the model's parameters. Examine the effect on the resulting shape of the airfoil caused by relocating the circle.
2. Create a model that generates the streamlines of Figure 5.50. These represent the air flow around the airfoil. This very challenging activity can be produced by using a generalization of the technique of Section 5.6.

Figure 5.50. Streamlines.

178 *Ovals of Cassini*

5.8. OVALS OF CASSINI

The complex square root function $f(z) = z^{1/2}$ can be used to construct a classical family of curves, the *ovals of Cassini*. This is done by creating a circle of radius r centered at $(h,0)$ on the real axis, and then evaluating $f(z) = z^{1/2}$ at each point on the circle. Because there are two square roots of each number, some creativity is required to compute the correct one. The family contains three types of curves, as discussed below and shown with $h = 0.95$ in Figures 5.51 to 5.53.

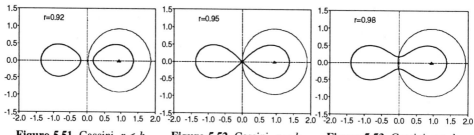

Figure 5.51. Cassini, $r < h$.   **Figure 5.52.** Cassini, $r = h$.   **Figure 5.53.** Cassini, $r > h$.

The model in Figure 5.54 creates the graph. Cells B2 and D2 are used to set r and h, with the constant $\alpha = \pi/72$ set in Cell F2. The base circle is then created using the standard technique by generating 144 points in Rows 4..148 of Columns C..D. However, because the arctangent function is continuous on the interval $-\pi \leq \theta \leq \pi$, but not on $0 \leq \theta \leq 2\pi$, the values of θ are designed to start at $-\pi$ in Cell B4. The polar coordinates r,θ for these points $z = x+iy$ are computed as usual in Columns E..F to give $z = x+iy = re^{i\theta}$.

```
         A     B       C      D       E      F       G       H
     1  Ovals of Cassini
     2   r=   0.98  h  =  0.95       α  =  0.044
     3   n    θ      x      y1       r      θ       y2      y3
     4   0   -3.14  -0.03  -0.00    0.03   -3.14
     5   1   -3.10  -0.03  -0.04    0.05   -2.17
     :
   148  144   3.14  -0.03   0.00    0.03    3.14
   149   0            0.00                         -0.17
   150   1            0.11                         -0.20
     :
   293  144            0.00                          0.17
   294   0           -0.00                                  0.17
   295   1           -0.11                                  0.20
     :
   438  144           -0.00                                 -0.17

F2:   @PI/72
A4:   0    A5:  1+A4                    B4:  -@PI    B5:  +B4+F$2
C4:   +B$2*@COS(B4)+D$2                 D4:  +B$2*@SIN(B4)
E4:   @SQRT(C4^2+D4^2)                  F4:  @ATAN2(C4,D4)
C149: @SQRT(E4)*@COS(F4/2)              A149: +A4
G149: @SQRT(E4)*@SIN(F4/2)              A294: +A4
C294: @SQRT(E4)*@COS(F4/2+@PI)
H294: @SQRT(E4)*@SIN(F4/2+@PI)
```

Figure 5.54. Ovals of Cassini.

Copy:		Graph:			
From	To	Series	Cells	Labels	Purpose
A5..B5	A6..B148	X	C4..C438		x-axis
C4..F4	C5..F148	1	D4..D148		circle
A149..G149	A150..G293	2	G4..G293		f(z)
A294..H294	A295..H438	3	H4..H438		f(z+π)

Next, to create the image, the square root of the initial point z is calculated in Cells C149 and G149 as $r^{1/2}e^{i\theta/2}$. The formulas in these cells are copied down through Row 293. Now the other square root must be generated. This is done in Cells C294 and H294 by calculating the alternate square root for the initial point z as $r^{1/2}e^{i\theta/2+\pi}$. The output curve is completed by copying these formulas down through Row 438. The graphs are produced as indicated in the table.

Once the model has been created, three classes of ovals are generated by varying r and h. The cases $r < h$, $r = h$, and $r > h$ produce a pair of ovals, a lemniscate, and a single curve, respectively.

Activities

1. Create the model of Figure 5.54 and experiment with its parameters.
2. Modify the model of Figure 5.54 to generate the ovals from a circle whose center is at the point (h,k), with k not necessarily 0. A new problem of continuity must now be considered.

5.10. INVERSES OF COMPLEX FUNCTIONS VIA SPREADSHEET SOLVERS

It is possible to modify the solver models of Section 2.15 to create representations of functions of a complex variable in a different fashion. In this approach, a curve C is specified as the image in the range of a function f, and the solver is used to find its pre-image C' in the domain. Thus, this process finds a representation for the inverse of a function without the need to solve for it algebraically. This is particularly important because functions in many applications of complex analysis arise in this fashion.

In the model of Figure 5.55, the values for θ, r_0, and r are entered in Columns C..E as in Section 2.15, with the (x,y)-values for the curve C' in the domain determined from these variables in Columns F..G by $x = a + r\cos(\theta)$ and $y = b + r\sin(\theta)$. The coordinates of the center point (a,b) for the solver technique, as described in Section 2.15, are set in Cells D4 and D5.

The values of the corresponding points $f(z) = u + iv$ of the image curve are determined in Columns H..I. In the example of Figure 5.57 the function is $f(z) = z^2 = (x^2-y^2) + 2xyi$, so $u = x^2-y^2$ and $v = 2xy$. Finally, the formula that defines the desired image curve C is entered in Column J. In this example, the image curve is a circle, $(u-x_0)^2 + (v-y_0)^2 = R^2$. The coordinates of (x_0,y_0) are set in Cells F4..F5 and the value of R^2 is entered in Cell H4. The solver command is used by forcing $(u-x_0)^2 + (v-y_0)^2$ to be set to R^2 in Column J.

180 *Inverses of Complex Functions via Spreadsheet Solvers*

As in Section 2.15, an initial estimate for the r_0 values of points in the pre-image is entered in Cell D12 and copied down Column D. Then the solver command is used to adjust the values for r_0 in Column D, and thereby each pair (x,y) and (u,v), to force the value in Column J to become R^2. When this occurs, the (u,v) values in Columns H..I lie on the image curve C. Because $u + vi = f(z)$, where $z = x + yi$ is given in Columns F..G, the points (x,y) lie on the pre-image C'. Thus, Columns F..G provide the graph of C', while Columns H..I provide the curve C. Both graphs can be displayed simultaneously on the screen. The graphs of Figures 5.56 and 5.57 are produced by the model.

As curves are produced, often it can be observed that there is more than one pre-image for a given point. As a result, sometimes the image is traced twice as the pre-image is traced out. At other times two or more distinct curves serve as the pre-image of a given curve C. These curves can be found by varying the initial value of r entered down Column D (see Exercise 2).

```
        B       C       D       E       F       G       H       I       J
  3   Solv.Center:     Circle:         Constant:
  4           a =     0   x0 =   1.5  Goal,R²=    3
  5           b =     0   y0 =     0  pi/72=   0.04
  :
 10                                 Input Curve    Output Curve
 11       n   θ:rad   r0      r     x       y       u       v       f
 12       0   0.00   1.80   1.80   1.80    0.00    3.23    0.01    3.00
 13       1   0.04   1.79   1.79   1.79    0.08    3.21    0.29    3.00
 14       2   0.09   1.79   1.79   1.78    0.16    3.14    0.56    3.00
  :
156     144   6.28   1.80   1.80   1.80    0.00    3.23    0.01    3.00

H5:  @PI/72                          B12: 0    B13: 1+B12
C12: +B12*H$5+0.001                  E12: @ABS(D12)
F12: +E12*@COS(C12)+D$4              G12: +E12*@SIN(C12)+D$5
H12: +F12^2-G12^2                    I12: 2*F12*G12
J12: (H12-F$4)^2+(I12-F$5)^2
```

Figure 5.55. Pre-Image of Complex Function via Solver.

It is easy to modify the model for another function f and image curve C. Figure 5.58 provides the pre-image of the curve $|x| + |y| = 2$ of Figure 5.59, produced from the function $f(z) = z^2$. Figures 5.60 is the pre-image of the circle in Figure 5.61 from $f(z) = z + 1/z$.

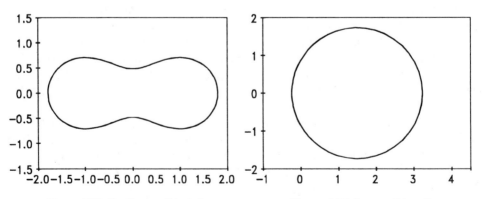

Figure 5.56. Pre-Image: $f(z) = z^2$.

Figure 5.57. Image: $f(z) = z^2$.

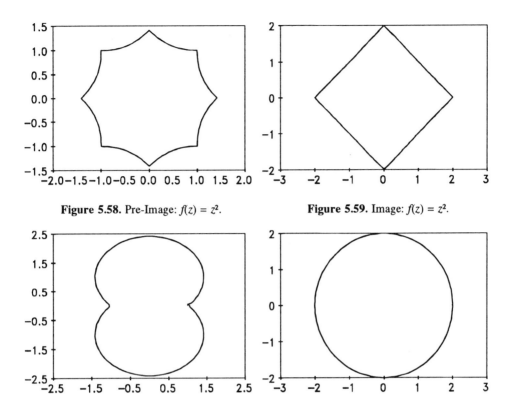

Figure 5.58. Pre-Image: $f(z) = z^2$.

Figure 5.59. Image: $f(z) = z^2$.

Figure 5.60. Pre-Image: $f(z) = z + 1/z$.

Figure 5.61. Image: $f(z) = z + 1/z$.

Activities

1. Create graphs for pre-images created from a variety of complex functions and image curves. Illustrative examples are listed in Kober (1957), with many in Spiegel (1964) and in the appendices of such complex variable books as Saff and Snider (1976), Fisher (1990), and Churchill (1960). Other examples that can be tried include:

   (a) $f(z) = z^2$, C: $(x^2)^{1/3} + (y^2)^{1/3} = (a^2)^{1/3}$ (astroid)
   (b) $f(z) = z^2$, C: $((1+x)^2)^{1/3} + (y^2)^{1/3} = 2$
   (c) $f(z) = z^2$, C: $(x^2+y^2)^2 - b(x^2-y^2) = 0$, with $b = 4$ (lemniscate)
   (d) $f(z) = z + 1/z$, C: $((1+x)^2)^{1/3} + ((2+y)^2)^{1/3} = 5$

2. Modify the model of Figure 5.55 to implement $f(z) = z + 1/z$ with the lemniscate $(x^2+y^2)^2 - b(x^2-y^2) = 0$, $b = 6$, as the image curve C. Show that two different pre-images are generated, by starting with $r = 0.5$ and $r = 2.0$.

3. Once a pre-image curve has been constructed, it can then be used with another function to produce a new image curve simply by changing the function used in Columns H..I. In this stage the solver is not used. Find the image curves produced with the pre-image curves generated above and using: (a) $f(z) = z^2$; (b) $f(z) = z^3$; (c) $f(z) = z + 1/z$; (d) $f(z) = e^z$; (e) $f(z) = \sin(z)$, where $f(x+iy) = 0.5(e^y - e^{-y})\cos(x) + 0.5i(e^y - e^{-y})\sin(x)$.

4. Create a model that finds pre-images of various curves with the function $f(z) = \sin(z)$. In particular, starting with an image consisting of a series of horizontal lines, create the channel streamline diagram of Figure 5.46. For the latter activity, a good location for the solver's central point is $(a,b) = (0,1.57)$, with an initial estimate of $r = 2$.

Appendix 1

SELECTED CURVES AND THEIR DERIVATIVES

Curve	Equation	Derivative, dy/dx
Archimedes' Spiral	$r = a\theta$	$[\sin(\theta)+\theta\cos(\theta)]/[\cos(\theta)-\theta\sin(\theta)]$
Astroid	$x = a\cos^3(\theta), y = a\sin^3(\theta)$	$-\tan(\theta)$
Bifolium	$r = a\sin(\theta)\cos^2(\theta)$	$[2\tan(\theta)\cos(2\theta)]/[2\cos(2\theta)-1]$
Cardioid	$r = 2a(1\pm\cos(\theta))$	$-[\cos(2\theta)\pm\cos(\theta)]/[\sin(2\theta)\pm\sin(\theta)]$
Circle	$r = a$; $x = a\cos(\theta), y = a\sin(\theta)$	$-x/y$, $-\cot(\theta)$
Cissoid of Diocles	$r = a(\csc(\theta)-\sin(\theta))$	$[\sin(3\theta)-3\sin(\theta)]/[\cos(3\theta)-3\cos(\theta)]$
Conchoid:Nicomedes	$r = b + a/\sin(\theta)$	$-b\cos(\theta)\sin^2(\theta)/(a+b\sin^3(\theta))$
Curtate Cycloid	$x = a\theta-b\sin(\theta),$ $y = a-b\cos(\theta), a>b$	$b\sin(\theta)/(a-b\cos(\theta))$
Cycloid	$x = a(\theta-\sin(\theta)), y = a(1-\cos(\theta))$	$\sin(\theta)/(1-\cos(\theta))$
Deltoid	$x = a(2\cos(\theta)+\cos(2\theta)),$ $y = a(2\sin(\theta)-\sin(2\theta))$	$[\cos(\theta)-1]/\sin(\theta)$
Ellipse	$x = a\cos(\theta), y = b\sin(\theta)$	$-[b^2x]/[a^2y]$; $-b/[a\tan(\theta)]$
Epicycloid	$x = a(n\cos(\theta)-\cos(n\theta)),$ $y = a(n\sin(\theta)-\sin(n\theta))$	$[\cos(n\theta)-\cos(\theta)]/[\sin(\theta)-\sin(n\theta)]$
Epitrochoid	$x = an\cos(\theta)-b\cos(n\theta),$ $y = an\sin(\theta)-b\sin(n\theta)$	$[b\cos(n\theta)-a\cos(\theta)]/[a\sin(\theta)-b\sin(n\theta)]$
Hyperbola	$x = a\sec(\theta), y = b\tan(\theta)$	$[b^2x]/[a^2y]$; $b/[a\sin(\theta)]$
Hyperbolic Spiral	$r = a/\theta$	$[\sin(\theta)-\theta\cos(\theta)]/[\cos(\theta)+\theta\sin(\theta)]$
Hypocycloid	$x = a(n\cos(\theta)+\cos(n\theta)),$ $y = a(n\sin(\theta)-\sin(n\theta))$	$[\cos(n\theta)-\cos(\theta)]/[\sin(n\theta)+\sin(\theta)]$
Hypotrochoid	$x = an\cos(\theta)+b\cos(n\theta),$ $y = an\sin(\theta)-b\sin(n\theta)$	$[b\cos(n\theta)-a\cos(\theta)]/[b\sin(n\theta)+a\sin(\theta)]$
Involute of Circle	$x = a(\cos(\theta)+\theta\sin(\theta)),$ $y = a(\sin(\theta)-\theta\cos(\theta))$	$\tan(\theta)$

Selected Curves and Their Derivatives

Curve	Equation	Derivative, dy/dx
Kampyle of Eudoxus	$r = a\sec^2(\theta)$	$[1+\sin^2(\theta)]/[\sin(\theta)\cos(\theta)]$
Lemniscate	$r^2 = a^2\cos(2\theta)$	$-\cot(3\theta)$
Limaçon	$r = 2a\cos(\theta) + b$	$-[2a\cos(2\theta)+b\cos(\theta)]/[2a\sin(2\theta)+b\sin(\theta)]$
Lissajous Curve	$x = a\sin(b\theta)$, $y = c\sin(d\theta)$	$[cd\cos(d\theta)]/[ab\cos(b\theta)]$
Lituus	$r^2 = a\theta$	$[\sin(\theta)+2\theta\cos(\theta)]/[\cos(\theta)-2\theta\sin(\theta)]$
Logarithmic Spiral	$r = ae^{b\theta}$	$[b\sin(\theta)+\cos(\theta)]/[b\cos(\theta)-\sin(\theta)]$
Nephroid	$x = a(3\cos(\theta)-\cos(3\theta))$, $y = a(3\sin(\theta)-\sin(3\theta))$	$[\cos(3\theta)-\cos(\theta)]/[\sin(\theta)-\sin(3\theta)]$
Parabola	$x = \tan(\theta)$, $y = a\tan^2(\theta)$	$2ax$; $2a\tan(\theta)$
Piriform	$x = a(1+\sin(\theta))$, $y = b\cos(\theta)(1+\sin(\theta))$	$[b(\cos(2\theta)-\sin(\theta))]/[a\cos(\theta)]$
Prolate Cycloid	$x = a\theta-b\sin(\theta)$, $y = a-b\cos(\theta)$, $a<b$	$[b\sin(\theta)]/[a-b\cos(\theta)]$
Right Strophoid	$r = a(\sec(\theta)-2\cos(\theta))$	$[\sec^2(\theta)-2\cos(2\theta)]/[2\sin(2\theta)]$
Rose	$r = \sin(n\theta)$	$[n\cos(n\theta)\sin(\theta)+\sin(n\theta)\cos(\theta)]/[n\cos(n\theta)\cos(\theta)-\sin(n\theta)\sin(\theta)]$
Semi-cubic Parabola	$r = \tan^2(\theta)\sec(\theta)/a$	$(3/2)\tan(\theta)$
Serpentine Curves	$x = a\cot(c\theta)$, $y = b\sin(d\theta)$	$-bd\cos(d\theta)\sin^2(c\theta)/[ac]$
Trisectrix(Maclaurin)	$r = a\sec(\theta) - 4a\cos(\theta)$	$[1+4\cos^2(\theta)-8\cos^4(\theta)]/[8\sin(\theta)\cos^3(\theta)]$
Tschirnhausen Cubic	$r = a\sec^3(\theta/3)$	$-\cot(2\theta/3)$
Witch of Agnesi	$x = 2a\cot(\theta)$, $y = 2a\sin^2(\theta)$	$-2\sin^3(\theta)\cos(\theta)$

Notes:

1. Frequently these curves can be reproduced in a different orientation by replacing functions by co-functions. For example, alternate forms are possible for the cardioid, $r = 2a(1\pm\sin(\theta))$ and the lemniscate, $r^2 = a^2\sin(2\theta)$.
2. Trigonometric identities can often be used to obtain alternate expressions. For example, another expression for the right strophoid is $r = a\cos(2\theta)\sec(\theta)$.

Appendix 2

GRAPHS IN WORD PROCESSING

The capability of incorporating a spreadsheet graph into a word processing document can be extremely advantageous in preparing documents. In fact, all of the illustrations in this book were first created using *Quattro Pro* and then included in a *Word Perfect* file. The process that was used is described below. Other combinations of spreadsheets and word processing programs are possible as well, and some other frequently used combinations are also mentioned.

A. *Quattro Pro* and *Word Perfect*

To use the combination of *Quattro Pro* and *Word Perfect*, a spreadsheet model that contains a graph is first created. Next, the graph is written to a file that can be included later in a word processing document. There are several types of graphic files that can be created. The following commands create a *.PIC file, which is given the name SAMPLE.PIC.

```
/ Print  Graph print  Write graph file
  PIC file  Enter file name: SAMPLE
  Quit  Quit  Quit
```

After this command is issued, a user leaves *Quattro Pro* and enters *Word Perfect*. Next, a *Word Perfect* document is either created or retrieved. In *Word Perfect*, the cursor is moved to the location within the document where the graph is to be included. Then one of the various types of graphics boxes, for example a User Box, is created. Finally, appropriate definitions are set, as shown in the example below.

```
Graphics, 4 User Box, 1 Create

1 File Name: SAMPLE.PIC
3 Caption: Figure 6.8. Sample Graph.
6 Horizontal Position: Center
7 Size: 3 Set Both  Width: 2.55  Height: 2.15
8 Wrap Text Around Box: No
9 Edit: 2 Scale  Scale X: 120  Scale Y: 120
```

In this example, first a User Box is created to contain the graph file SAMPLE.PIC (1). Next, a caption for the box is provided in the format used throughout this book (3), with the box centered on the page (6). The size of the box is set manually to the most commonly used size employed within this book (7). Also, the option is chosen by which the text is not wrapped around the box (8). Finally, an edit option (9) is used to adjust the relative size of the graph and its position within the box. After the box has been created in this fashion, additional blank lines are inserted beneath it in order to move the next line of text a sufficient distance below the box.

There are many possible alternatives beyond those selected above. The horizontal position (6) can also be set to Left or Right so that two boxes can be displayed on the same line. If desired, text can be wrapped around the graphics box (8). The image in the box can also be moved or rotated (9). There are also overall alternatives that can be set for each type of box, such as the style of surrounding borders and the size of the margins around the box. These can be set to define a standard convention at the beginning of a document through the Graphics, User Box, Options command. Details can be located in appropriate manuals.

Quattro Pro allows several types of graphic files to be created. Each of these has its own advantages and drawbacks, which depend somewhat on the capabilities of the printer and the computer being used, as well as with the desired application. Experimentation can provide valuable insight into determining an appropriate type.

The type chosen in option (2) of the example above was the *.PIC file. This is the type most frequently used in this book. When included in a *Word Perfect* document, the *.PIC files provide nice, smooth curves that can be printed on most printers. Moreover, they can also be examined in the screen preview mode. Unfortunately, *.PIC graphs usually do not preserve all features that are present on the screen. Such desirable features as dashed, dotted, or heavy lines are generally replaced by solid curves of a standard thickness.

Another type of graphic file that can be created is the encapsulated postscript file, with an extension of *.EPS. This type has been used occasionally in this book, for example in Figure 3.1. The *.EPS files produce high-quality graphic output and allow the display of dashed, dotted, and heavy lines. However, a postscript printer is required to print the ensuing graph, and the graphs cannot be viewed via the screen print preview option. Also, the resulting curves that are printed may contain an unexpected waviness.

A third type of graphic file is the *.PCX, or paintbrush file. This type also preserves most of the spreadsheet graph's features, but generally provides only fair graphic quality.

Quattro Pro also enables both the numerical spreadsheet display and the underlying formulas to be printed to a file through options under the Print command.

B. *Lotus 1-2-3* and *Word Perfect*

A procedure similar to the one above may be used with *Lotus 1-2-3*. Once a graph is created, the following commands create the graph file which can be used in *Word Perfect*, just as above. In Version 2.3 the file created by *Lotus* is SAMPLE.PIC, while with Version 3.0 it is SAMPLE.CGM. The resulting graph can be embedded in *Word Perfect* just as described for *Quattro Pro*.

```
/ Graph  Save
Enter name of file: SAMPLE
```

C. *Word Perfect*'s Grab Command

Another type of file can be obtained from a graphic display through the screen capture GRAB command that is provided by the *Word Perfect* program itself. Although a manual should be read for details, the general process is easy to describe. First, if the file GRAB.COM has been installed, the Grab command is activated from DOS by typing **GRAB**. A spreadsheet program such as *Quattro Pro* or *Lotus 1-2-3* is then opened. When a file that contains a graph is brought into memory, and the desired graph is displayed on the screen, the Grab command is issued by pressing the **<ALT-SHIFT-F9>** keys simultaneously. This causes a box to be superimposed upon the graph to indicate the part of the graph that will be selected. This box can be adjusted by using the arrow and shift keys. When the appropriate part of the graph is enclosed in the box, a graphics file is created by pressing the **<ENTER>** key. The resulting file is supplied with the name GRAB.WPG. This file can be renamed subsequently by the DOS rename command. The graph that is produced typically is of a reasonably good quality.

D. *Excel* and *Microsoft Word*

Another popular combination of spreadsheet and word processor is of *Microsoft Excel* and *Word*. The process used with this pair is somewhat different from the one above. Both programs are run under *Microsoft Windows*. To imbed a graph into *Word*, first *Word* and then *Excel* are opened. While an *Excel* file that includes an embedded graph is the active file, the mouse is used to select the graph. Then the **<SHIFT-EDIT>** combination is pressed to produce a special edit menu. From this menu one selects the Copy Picture option and appropriate sub-selections that determine the nature of the appearance of the graph when it is printed. Next, a user switches to the *Word* program, and sets the cursor at the position within the *Word* document where the graph is to be located. Once in *Word*, the Edit option is chosen, followed by the selection of Paste Special. From the subsequent menu choices that are presented, selecting the Paste Picture option causes the graph to be included in the *Word* document as a picture. The key steps are summarized below.

Excel
select graph
<SHIFT>-Edit
Copy Picture
select options

Word
Edit
Paste Special
Paste Picture

As in Section A above, it is also possible to embed spreadsheet numerical output and formulas from *Excel* into the *Word* file.

Appendix 3

DISK OF EXAMPLES

This book is accompanied by a disk that includes 25 illustrative examples selected from throughout the book and created in each of the spreadsheets *Quattro Pro*, *Microsoft Excel*, and *Lotus 1-2-3*. The one exception is that there is no solver example (Figure 2.103) for *Lotus*. These examples, which are listed in Figure A.1, are stored on the disk in the subdirectories, QUATTRO, EXCEL, and LOTUS. Each file name is devised to suggest the related model. The final digit of a name indicates the chapter in which it is found. The *Quattro Pro* files were created using Version 2.0, the *Lotus 1-2-3* files using Version 2.3, and the *Microsoft Excel* files using Version 4.0. Each set of files should be usable with those versions as well as higher numbered ones. The *Lotus* files also are read by Version 2.2. The disk has a double density format that should be accepted by most computers.

Figure	Example	Description
2.5	PARAM2	Parametric/Polar Equations
2.18	CARD_EN2	Cardioid as Envelope of Circles
2.20	CARD_IT2	Multiple Curves for Parameter a
2.28	DURER2	Dürer's Construction of Limaçon
2.35	AST_LIN2	Astroid via Lines (Nail and String)
2.37	AST_ELL2	Astroid as Envelope of Ellipses
2.40	AST_X2	Astroid as Envelope of Lines (Alt)
2.66	LISSAJ2	Lissajous Curves
2.81	CYCLOID2	Cycloid via Series of Circles
2.97	ROSE2	Rose Model
2.103*	SOLVER2	Lemniscate via Solver Command (not Lotus)
3.2	STROPH3	Strophoid Construction
3.19	PEDAL3	Pedal Construction
3.27	NEG_PED3	Negative Pedal Envelope via Lines
3.33	INVERSE3	Construction of Inverse Curve
3.37	PARALLE3	Construction of Parallel Curve
4.3	TRACE4	Animated Trace of Parametric Curve
4.8	VARY4	Effect of Varying a Parameter
4.15	ELL_CIR4	Ellipse via Circles and Distances
4.36	CYCLOID4	Circle Rolls on Line Forming Cycloid
4.44	HYPOCYC4	Circle Rolls in Circle Forming Hypocycloid
4.63	LINKAGE4	Animated Linkage Construction
5.14	CPL_EXP5	Representation of Complex Function
5.34	EX_GRID5	Grid Representation of Complex Function
5.45	STREAM5	Fluid Flow Streamlines

Figure A.1. Disk Examples.

190 *Disk of Examples*

In order to store a large number of large files on one double density disk, each of the examples has been condensed by using the spreadsheets' delete commands to delete most of the rows from the large, iterative blocks of each model. It is easy to restore these examples to their full size on a computer by using the standard insert and copy routine described in Chapter 1. The act of carrying out this procedure reinforces a useful spreadsheet technique. However, in order to facilitate and accelerate the process, each example also contains a built-in macro command to do the expansion automatically.

One procedure to use in expanding the files on the disk is described below. Figure A.2 shows Columns A..I of Rows 10..23 of the condensed *Quattro Pro* disk file for the cardioid envelope model of Figure 2.18. It is stored in the subdirectory QUATTRO. In the form in which it is stored on the disk, the model will produce only the first few points of a curve.

	A	B	C	D	E	F	G	H	I
10		Curve:					Curve	Dist	Aux
11		n	θ(deg)	θ(rad)	r	x	y1	r	y2
12		0	0	0.00		1.00	0.00	0.00	
13		1	15	0.26		0.97	0.26	0.26	
14		2	30	0.52		0.87	0.50	0.52	
15	36	3	45	0.79		0.71	0.71	0.77	
16		Additional:							
17		Fixed Point:				1			
18		Equation:				999			
19		Circles:							
20		1	0	0.0		1.23			0.26
21		1	1	0.3		1.22			0.33
22		1	2	0.5		1.19			0.39
23	663	1	3	0.8		1.15			0.44

Figure A.2. Layout of Disk File.

The most efficient way to use the example files is to transfer them first to a hard disk drive. However, they can also be used directly from the disk provided. In either case, a user first loads the spreadsheet program, and then chooses the directory that contains the files for that spreadsheet, here *Quattro Pro*. Next, the file CARD_EN2.WQ1 is opened, just as described in Chapter 1. Following this, the file is expanded by the insert and copy procedure.

In nearly all of the disk examples, Column A is used to indicate the number of the row in which the last line of a block will ultimately reside. In a few examples, primarily those from Chapter 5, Column A has already been used for other purposes. In these examples the corresponding row numbers are shown in a column along the right side of the model. The basic approach used for the expansion procedure is to insert additional rows immediately in front of the last row in each block.

To modify this example, first move the cursor to Cell B15. Then insert new, additional rows in front of this row and in the process move the current Row 15 to Row 36. This is done by inserting rows through Row 35. Next, copy the contents of the penultimate row of the old block, Row 14, through the new terminal row, Row 36. The following commands can be used in *Quattro Pro* to complete the first block, and thereby generate the base circle for the particular model.

> / E(dit) I(nsert) R(ows) Rows: **B15..B36**
> / E(dit) C(opy) Source: **B14..H14** Destination: **B15..B36**

Next, expand the second block of the model in the same way. This block generates the family of circles that forms the envelope. To do this, move the cursor to the row that contains the value 663 in Column A. After the steps above have been carried out, this will be Row 44. Now repeat the insert and copy procedure using the commands below.

> / E(dit) I(nsert) R(ows) Rows: **B44..B662**
> / E(dit) C(opy) Source: **B43..I43** Destination: **B44..G663**

Finally, move the cursor to an appropriate cell at the top of the spreadsheet. The resulting file can now be saved in a form identical to that of Figure 2.18. For the most part, each of the files on the disk has been created just as described in the text. Again, it should be noted that these models were designed to be small enough to be accommodated on many smaller computer configurations. It may be very desirable to enlarge a model further by decreasing a step size and increasing the number of points employed in the example by additional use of the insert and copy technique in order to produce even smoother curves. Also, the models' layouts have not been embellished with additional user instructions or format changes, so that the disk example will be the same as the book's descriptions. Interested users are encouraged to create these refinements themselves.

For individuals who do not want to continually repeat the insert and copy procedure manually, each file contains a macro which carries out the expansion procedure automatically. In *Quattro Pro* and *Lotus 1-2-3* these macros are named as \A, and they are stored in the upper right corner of each model. They are implemented by pressing the combination <ALT-A>. For example, Figure A.3 shows the macro for the example of Figure A.1. The macro contains the key strokes for the procedures discussed above.

```
{GOTO}B15~
/EIRB15..B35~
/ECB14..H14~B15..B36~
{GOTO}B44~
/EIRB44..B662~
/ECB43..I43~B44..B663~
{GOTO}B1~{GOTO}G3~
```

Figure A.3. Expansion Macro.

Some versions of *Lotus 1-2-3* provide an "undo" feature. Because this feature is quite demanding of memory capacity, it should be disabled before executing the expansion macros. Otherwise, an "inadequate memory" error may result. This feature is disabled by issuing the following series of commands: /Worksheet, Global, Default, Other, Undo, Disable, Quit.

Most of the *Microsoft Excel* macros are stored in a separate macro sheet, MACS.XLM, with a goal-seek macro contained in the file SEEKMAC.XLM. Once an *Excel* spreadsheet file on the disk is opened, the related macro is implemented by opening the file MACS.XLM or SEEKMAC.XLM and then selecting the desired macro. For this example, the mouse can be used to choose the following options:

> Macro Run **card_en2**

Some graphs require the use of the PENUP macro on the file MACS.XLM to erase the contents of spreadsheet cells that contain NA. This implements the pen-up capability discussed in Section 1.8.

In using the disk files, there are a few things to note. First, the *Quattro Pro* solver model for Figure 2.103 employs two macros. The first one, \A, expands the model, just as discussed above. The second one, \G, carries out the solver macro procedure that is presented in Section 2.15. The corresponding *Excel* file for Figure 2.103, SOLVER2.XLS, is stored in its full form, so there is no need to use an expansion macro with it. In addition, rather than employing *Excel*'s more complex *Excel* solver command, the GOAL.SEEK command has been used in the macro Seek that is stored in the file SEEKMAC.XLM and shown below.

```
A2: S= 12
A3: =GOAL.SEEK("R"&TEXT(S,0)&"C8",1.21,"R"&TEXT(S,0)&"C4")
A4: S= S+1
A5: =IF(S<84,GOTO(A3))
A6: =RETURN()
```

Figure A.4. Excel Goal.Seek Macro

Before using this macro, a user must enter a constant for the target value of k^2 (here 1.21) that appears in Cell G4 of the spreadsheet model by editing the entry in Cell A3 of the macro sheet. At present, *Excel* does not allow the use of a FOR..NEXT loop with Solver or Goal.Seek macros, so the equivalent GOTO structure has been used. Writing a similar macro using *Excel*'s solver commands is a good exercise.

In *Quattro Pro* and *Lotus 1-2-3* files, the graphs that have been created are viewed with / Graph View command. The graphs also can be imbedded in the spreadsheet by users. In order to save space, the disk's *Excel* examples contain graphs that are embedded at the bottom of each model. Although this is probably not where one ultimately wants them placed, this location facilitates the file expansion process. After the model is expanded, the graphs can be repositioned to more advantageous locations through the use of a mouse.

REFERENCES

Althoen, S. C. and Wyneken, M. F., The width of a rose petal, *Am. Math. Monthly*, 97(10), 907–911, December 1990.

Anon. Harvard University sophomore makes straight-line drawings, *Life*, 8(12), 43–44, March 18, 1940.

Arganbright, D. E., *Mathematical Applications of Electronic Spreadsheets*, McGraw-Hill Book Co., New York, 1985.

Arganbright, D. E., Using spreadsheets in teaching statistics, *Statistics for the Twenty-First Century*, The Mathematical Association of America, Washington, DC, 226–242.

Beyer, W. H, Ed., *CRC Standard Mathematical Tables and Formulas*, 29th ed., CRC Press, Boca Raton FL, 1992.

Bolt, B., *Mathematics Meets Technology*, Cambridge University Press, Cambridge, 1991.

Brieskorn, E. and Knörrer, H., *Plane Algebraic Curves*, Birkhäuser Verlag, Basel, 1986.

Bruce, J. W. and Giblin, P. J., *Curves and Singularities*, 2nd ed., Cambridge University Press, Cambridge, 1992.

Bruce, J. W., Giblin, P. J., and Rippon, R. J., *Microcomputers and Mathematics*, Cambridge University Press, Cambridge, 1990.

Carter, C. R., Using technology in graphing, *Math. Teacher*, 85(2), 118–121, February 1992.

Churchill, R. V., *Complex Variables and Applications*, 2nd ed., McGraw-Hill Book Co., New York, 1960.

Courant, R. and Robbins, H., *What is Mathematics?* Oxford University Press, London, 1941.

Coxeter, H. S. M., *Introduction to Geometry*, 2nd ed., John Wiley & Sons, Inc., New York, 1969.

Cundy, H. M. and Rollett, A. P., *Mathematical Models*, Oxford University Press, London, 1961.

Fay, T. H., The butterfly curve, *Am. Math. Monthly*, 96(5), 442–443, May 1989.

Finney, R. L. and Thomas, G. B. Jr., *Calculus*, Addison-Wesley Publishing Co., Reading, MA, 1990.

Fisher, S. D., *Complex Variables*, 2nd. ed., Wadsworth and Brooks/Cole, Belmont, CA, 1990.

Frost, P., *An Elementary Treatice on Curve Tracing*, 5th ed., Chelsea Publishing Co., New York, 1960.

Greenleaf, F. P., *Introduction to Complex Variables*, W. B. Saunders Co., Philadelphia, 1972.

Gray, A., *Modern Differential Geometry of Curves and Surfaces*, CRC Press, Boca Raton, FL, 1993.

Hall, L. M., Trochoids, roses and thorns — beyond the spirograph, *Coll. Math. J.*, 23(1), 20–35, January 1992.

Hall, L. and Wagon, S., Roads and wheels, *Math. Mag.*, 65(5), 283–301, December 1992.

Hardin, D. and Strang, G., A thousand points of light, *Coll. Math. J.*, 21(5), 406–409, November 1990.

Hill, F. S., *Computer Graphics*, Macmillan Publishing Co., New York, 1990.

Hilton, H., *Plane Algebraic Curves*, 2nd ed., Oxford University Press, London, 1932.

Honsberger, R., *Mathematical Gems*, The Mathematical Association of America, Washington, DC, 1973.

Kober, H., *Dictionary of Conformal Representations*, Dover Publications, Inc., New York, 1957.

Krause, E. F., *Taxicab Geometry*, Dover Publications Inc., New York, 1986.

Larson, R. E., Kanold, T. D., and Stiff, L., Discovering life in hard data, *Math. Teacher*, 86(1), 8, January 1992.

Lawrence, J. D., *A Catalog of Special Plane Curves*, Dover Publications, Inc., New York, 1972.

Lockwood, E. H., *A Book of Curves*, Cambridge University Press, London, 1961.

Markushevich, A. I., *Remarkable Curves*, Mir Publishers, Moscow, 1980.

Maurer, P. M., A rose is a rose, *Am. Math. Monthly*, 94(6), 631–645, June-July 1987.

Millington, J., *Curve Stitching*, Tarquin Publications, Stradbroke, England, 1989.

Pedoe, D., *Geometry and the Visual Arts*, Dover Publications, Inc., New York, 1976.

Saff, E. B. and Snider, A. D., *Fundamentals of Complex Analysis*, Prentice-Hall, Inc., Englewood Cliffs, NJ, 1976.

Salmon, G., *Higher Plane Curves*, 3rd ed., Chelsea Publishing Co., New York, 1960.

Seymour, D., *Introduction to Line Designs*, Dale Seymour Publications, Palo Alto, CA, 1992.
Spiegel, M. R., *Theory and Problems of Complex Variables*, McGraw-Hill, New York, 1964.
Steinhaus, H., *Mathematical Snapshots*, 3rd Am. ed., Oxford University Press, Oxford, 1969.
Stewart, J., *Calculus*, 2nd ed., Brooks/Cole Publishing Company, Belmont, CA, 1991.
Taylor, W. F., *The Geometry of Computer Graphics*, Wadsworth & Brooks/Cole, Pacific Grove, CA, 1992.
Teixeira, F. G., *Traite des Courbes, Tomes I-III*, Chelsea Publishing Co., New York, 1971.
Von Seggern, D., *CRC Standard Curves and Surfaces*, 2nd ed., CRC Press, Boca Raton, FL, 1993.
Wells, D., *The Penguin Dictionary of Curious and Interesting Geometry*, Penguin Books, London, 1991.
Yaglom, I. M. and Boltyanskii, V. G., *Convex Figures*, Holt, Rinehart and Winston, New York, 1961.
Yates, R. C., *Curves and Their Properties*, The National Council of Teachers of Mathematics, Reston, VA, 1974.

INDEX

absolute reference, 3, 13
affine transformations, 169
Agnesi, witch of, 73–74
airfoil, 176-177
algebraic curves, 81–83
animation, 117–154
approximation model, 113–116
Archimedes's spiral, 65
argument, 156
astroid, 45–51, 97, 101–102, 106, 181
auxiliary curves, 18–20

bicorn, 84
bifolium, 74, 95
binomial distribution, 8
Birkhoff, George, 115
Bowditch figures, 63
bugs on table, 144
built-in function, 3
butterfly, 32, 35

cardioid, 32–34, 36–42, 88, 91, 97, 100, 109, 113, 119–121
carpenter's square, 146
Cartesian coordinates, 8
Cassini, ovals of, 178–179
caustic, 109–113
channel flow, 174–176
choose function, 23
circle, 38–40, 43, 52, 54–55, 59–60, 61, 68–70, 73, 85–88, 97–100, 109–111, 121–124, 132–135, 139–141, 163, 166–167, 176–177, 180–181
circle of inversion, 100
circular functions, 149–151
circular reference, 6
cissoid, 73, 92–94
cocked hat, 84
complex arithmetic, 155–160
complex functions, 161–172, 179–182
complex powers, 155–158
complex roots, 159–160
complex variables, 155–181
concatenation, 20
conchoid, 73–74, 89–91
conformal, 173

conic sections, 61–63, 122–132
conjugate, 155
constant width, 151–154
coordinates
   Cartesian, 8
   polar, 8, 32
copy command, 3
curtate cycloid, 71–72, 135
curve equations, 183–184
curve proportions, 15–16
curve sketching, 12–16, 75–84, 117–119
curve stitching, 47, 51
curves of constant width, 151–154
cycloid, 67–70, 132–135
   curtate, 71–72, 135
   prolate, 71–72, 135

delete command, 5–6
delta curve, 154
deltoid, 50, 52–53, 95–96, 103
derivatives, 183–184
dilation, 169
directrix, 125, 127
distance, 9
dot product, 10, 12
drawings, 115–116, 168
dumbbell, 84
Dürer's method, 43–44

eccentricity, 62, 121
eight curve, 32
electric motor, 84
ellipse, 38, 42, 48–49, 62, 92, 94, 98, 103–105, 107–108, 122–124, 127–130, 142
   taxicab, 83
embedding graphs, 16, 25, 26
envelope, 12
envelope models, 38–39, 42–45, 47–50, 53, 54–57, 63, 98–100, 106–113
epi spiral, 67
epicycloid, 70–71, 73, 134–135, 138–140
epitrochoid, 71–72
EPS files, 186
equiangular spiral, 66
evolute, 106–109

Fibonacci, numbers 8
fluid flow, 173–176
foci, 77, 122, 125, 127
four-leafed rose, 117

gear train, 138
geometry, taxicab, 83–84
geometric transformations, 168
glissettes, 142–144
goal-seek, 191
grab command, 186
graph command, 4
graphing, 4–5, 12–24, 161–163
grid representations, 169–172

head of vector, 9
hyperbola, 59–60, 61, 125–127, 130–131
   taxicab, 83
hyperbolic spiral, 66
hypocycloid, 71–73, 135–139
hypotrochoid, 72

insert command, 5
insert and copy procedure, 5
interest example, 2–3
interior labels, 14
inverse curves, 100–103
inverse of complex function, 179–181
inverse of real function, 2–3
involute of circle, 73, 139–141

Joukowski airfoil, 176–177

Lamé curve, 107
lemniscate, 57–61, 77–80, 84, 146–149,
   178–179, 181
length, vector, 9
library function, 3
limaçon, 42–45, 94, 103, 149
line drawing, 115–116
linear fractional transformation, 169
linear transformation, 169
lines, 10–11, 18–24, 47–51, 53, 56,
   87–88, 91, 98–100, 106–116, 144
linkages, 146–149
Lissajous figures, 63–65
lituus, 66–67
logarithmic spiral, 66
logistic function, 121

lookup function, 4
Lotus 1-2-3 topics, 24–25, 186

macro, 27, 79–80, 189–192
Maltese cross, 84
Microsoft Excel topics, 26–27, 187
Microsoft Word, 187
Middleton, David, 115
modulo function, 12
modulus, 156
multiple curves, 17–18, 40–41
multiple lines, 21–23

name command, 4
negative pedals, 98–100
nephroid, 52, 54–57
normal curve, 121
normal line, 11, 113–114
normal vectors, 10

ovals of Cassini, 178–179

parabola, 51, 62, 91, 100, 106, 109, 125,
   127, 130–131, 144
   taxicab, 84
parabolic spiral, 66–67
paraboloid, semicubical, 74
parallel curves, 103–106
parameters, varying, 119–121
parametric equations, 30–35
partial sums, 8, 158
PCX files, 186
pedal curves, 95–97
pencil and string, 127–130
pen-up, 22, 27
perpendicular vectors, 10
PIC files, 185–186
piriform, 73–74
Poinsot spiral, 35, 67
polar coordinates, 8, 32, 36–37
polar graphs, 36–37
polar grid, 36
primitive root, 159
prolate cycloid, 71, 135
powers, complex, 155–158

quadratrix, 131–132
Quattro Pro, 1–7, 185–186

ratio aspect, 15–16, 25
recalculation, 6
reciprocal spiral, 66
reference
  absolute, 3, 13
  circular, 6
  relative, 3, 13
relative reference, 3, 13
retrieve command, 4
resonance, 121
Reuleaux triangle, 151, 153
roots
  complex, 159–160
  primitive, 159
rose, 35, 74–75, 106, 117, 138
rotation, 62–63, 169

save command, 4, 5
savings account example, 2–3
scalar multiple, 10
sector of circle, 164–166
semicubical paraboloid, 74
series, infinite, 8, 158
serpentine, 65
Simpson's line, 53
slope, 10–11, 19
solver command, 75–84, 179–181
spiral, 65–67
  Archimedes, 65
  epi, 67
  hyperbolic (reciprocal), 66
  lituus, 66–67
  logarithmic (equiangular), 66
  parabolic, 66–67
  Poinsot, 35, 67
  square root, 67
spreadsheet graphs, 4–5, 12–24, 161–163

spreadsheet operation, 1–8
square root spiral, 67
stirrup, 84
straight-edge and triangle, 144–146
streamlines, 173–176, 177
string-and-nail design, 47
string function, 20
strophoid, 85
sum of vectors, 10
swallowtail, 32

tail, vector, 9
tangent line, 10, 18, 42, 111–114
taxicab geometry, 83–84
Taylor polynomials, 18
teardrop, 35, 42
tractrix, 116
transformation
  affine, 169
  geometric, 168
  linear, 169
  linear fractional, 169
triangle, Reuleaux, 151, 153
trigonometric functions, 121, 149–151
trigonometry, 11
Tschirnhausen's cubic, 73–74

unit vector, 10

varying parameters, 119–121
vector, 9–10

witch of Agnesi, 73–74
Word Perfect, 185–187
word processing, 185–187

xy graph, 4, 13